Biodiversity: An Introduction

Biodiversity:
An Introduction

Edited by
Jase Fitzgerald

Larsen & Keller
www.larsen-keller.com

Biodiversity: An Introduction
Edited by Jase Fitzgerald
ISBN: 978-1-63549-042-8 (Hardback)

📖 Larsen & Keller

Published by Larsen and Keller Education,
5 Penn Plaza,
19th Floor,
New York, NY 10001, USA

Cataloging-in-Publication Data

Biodiversity : an introduction / edited by Jase Fitzgerald.
   p. cm.
Includes bibliographical references and index.
ISBN 978-1-63549-042-8
1. Biodiversity. 2. Species diversity. 3. Biodiversity conservation. I. Fitzerald, Jase.
QH541.15.B56 B56 2017
577--dc23

The publisher's policy is to use permanent paper from mills that operate a sustainable forestry policy. Furthermore, the publisher ensures that the text paper and cover boards used have met acceptable environmental accreditation standards.

Printed and bound in the United States of America.

For more information regarding Larsen and Keller Education and its products, please visit the publisher's website www.larsen-keller.com

Table of Contents

Preface

This book provides comprehensive insights into the field of biodiversity. It attempts to understand the multiple branches that fall under this discipline and how such concepts have practical applications. Biodiversity refers to the various species of animals, plants, insects that are present on the different ecosystems on Earth. It includes topics like species variation, genetic variation and ecosystem variation. Biodiversity is an upcoming field of science that has undergone rapid development over the past few decades. The different types of biodiversity are terrestrial and marine biodiversity. This book will give detailed information about both of the above mentioned topics along with other ones. The textbook aims to serve as a resource guide for students and contribute to the growth of the discipline.

A detailed account of the significant topics covered in this book is provided below:

Chapter 1- There exist several species of flora and fauna on this Earth. Biodiversity refers to the numerous species, ecosystems, habitats and the genetic variations that exist in organisms. This chapter is an overview of the subject matter incorporating all the major aspects of biodiversity. There is a section dedicated to the topic of ecosystem as well.

Chapter 2- The chapter strategically encompasses and incorporates the major components and key concepts of biodiversity and provides a complete understanding of them. The topics discussed in this chapter are alpha diversity, extinction, defaunation, extinction event, snowball Earth, endemism and biodiversity hotspot. This chapter is a compilation of the various branches of biodiversity that form an integral part of the broader subject matter.

Chapter 3- Species can be classified according to their geological availability and the threat that humans pose to them. The chapter discusses the various types of species classification like rare, threatened, extinct, keystone, indicator, umbrella, flagship, introduced and invasive species. This chapter emphasizes about the need for preservation and conservation of species while also discussing the role that different species play in maintaining ecological balance.

Chapter 4- Human beings are one of the biggest threats to biodiversity. Human activity has led to habitat destruction, extinction of several species, pollution, climate change, overexploitation, human overpopulation etc. This chapter discusses the threats to biodiversity with special reference to these topics. One is able to understand the impact humans have on the environment by their day-to-day activities.

Chapter 5- To mitigate the damage to Earth's biodiversity, local governments have taken counter measures like the opening of animal sanctuaries, national parks, biodiversity banking, biodiversity offsetting and mitigation banking. These measures have begun a movement in the right direction and in this chapter readers are informed about how these steps have been carried out.

Chapter 6- On a global scale, there are several organizations in place to help safeguard biodiversity. There are also many treaties and agreements in place like the Convention on Biological Diversity, Cartagena protocol on Biosafety, Nagoya Protocol and the Biodiversity Indicators Partnership. This chapter comprehensively details all these initiatives and informs the reader about the objectives of each.

I would like to make a special mention of my publisher who considered me worthy of this opportunity and also supported me throughout the process. I would also like to thank the editing team at the back-end who extended their help whenever required.

Editor

Introduction to Biodiversity

There exist several species of flora and fauna on this Earth. Biodiversity refers to the numerous species, ecosystems, habitats and the genetic variations that exist in organisms. This chapter is an overview of the subject matter incorporating all the major aspects of biodiversity. There is a section dedicated to the topic of ecosystem as well.

Biodiversity, a contraction of "biological diversity," generally refers to the variety and variability of life on Earth. One of the most widely used definitions defines it in terms of the variability within species, between species and between ecosystems. It is a measure of the variety of organisms present in different ecosystems. This can refer to genetic variation, ecosystem variation, or species variation (number of species) within an area, biome, or planet. Terrestrial biodiversity tends to be greater near the equator, which seems to be the result of the warm climate and high primary productivity. Biodiversity is not distributed evenly on Earth. It is richest in the tropics. Marine biodiversity tends to be highest along coasts in the Western Pacific, where sea surface temperature is highest and in the mid-latitudinal band in all oceans. There are latitudinal gradients in species diversity. Biodiversity generally tends to cluster in hotspots, and has been increasing through time, but will be likely to slow in the future.

The number and variety of plants, animals and other organisms that exist is known as biodiversity. It is an essential component of nature and it ensures the survival of human species by providing food, fuel, shelter, medicines and other resources to mankind. The richness of biodiversity depends on the climatic conditions and area of the region. All species of plants taken together are known as flora and about 70,000 species of plants are known to date. All species of animals taken together are known as fauna which includes birds, mammals, fish, reptiles, insects, crustaceans, molluscs, etc.

Rapid environmental changes typically cause mass extinctions. More than 99 percent of all species, amounting to over five billion species, that ever lived on Earth are estimated to be extinct. Estimates on the number of Earth's current species range from 10 million to 14 million, of which about 1.2 million have been documented and over 86 percent have not yet been described. More recently, in May 2016, scientists reported that 1 trillion species are estimated to be on Earth currently with only one-thousandth of one percent described. The total amount of related DNA base pairs on Earth is estimated at 5.0×10^{37} and weighs 50 billion tonnes. In comparison, the total mass of the biosphere has been estimated to be as much as 4 TtC (trillion tons of carbon). In July 2016, scientists reported identifying a set of 355 genes from the Last Universal Common Ancestor (LUCA) of all organisms living on Earth.

The age of the Earth is about 4.54 billion years old. The earliest undisputed evidence of life on Earth dates at least from 3.5 billion years ago, during the Eoarchean Era after a geological crust started to solidify following the earlier molten Hadean Eon. There are microbial mat fossils found in 3.48 billion-year-old sandstone discovered in Western Australia. Other early physical evidence of a biogenic substance is graphite in 3.7 billion-year-old meta-sedimentary rocks discovered in Western Greenland. More recently, in 2015, "remains of biotic life" were found in 4.1 billion-year-old rocks in Western Australia. According to one of the researchers, "If life arose relatively quickly on Earth .. then it could be common in the universe."

Since life began on Earth, five major mass extinctions and several minor events have led to large and sudden drops in biodiversity. The Phanerozoic eon (the last 540 million years) marked a rapid growth in biodiversity via the Cambrian explosion—a period during which the majority of multi-cellular phyla first appeared. The next 400 million years included repeated, massive biodiversity losses classified as mass extinction events. In the Carboniferous, rainforest collapse led to a great loss of plant and animal life. The Permian–Triassic extinction event, 251 million years ago, was the worst; vertebrate recovery took 30 million years. The most recent, the Cretaceous–Paleogene extinction event, occurred 65 million years ago and has often attracted more attention than others because it resulted in the extinction of the dinosaurs.

The period since the emergence of humans has displayed an ongoing biodiversity reduction and an accompanying loss of genetic diversity. Named the Holocene extinction, the reduction is caused primarily by human impacts, particularly habitat destruction. Conversely, biodiversity impacts human health in a number of ways, both positively and negatively.

The United Nations designated 2011–2020 as the United Nations Decade on Biodiversity.

Etymology

The term biological diversity was used first by wildlife scientist and conservationist Raymond F. Dasmann in the year 1968 lay book *A Different Kind of Country* advocating conservation. The term was widely adopted only after more than a decade, when in the 1980s it came into common usage in science and environmental policy. Thomas Lovejoy, in the foreword to the book *Conservation Biology*, introduced the term to the scientific community. Until then the term "natural diversity" was common, introduced by The Science Division of The Nature Conservancy in an important 1975 study, "The Preservation of Natural Diversity." By the early 1980s TNC's Science program and its head, Robert E. Jenkins, Lovejoy and other leading conservation scientists at the time in America advocated the use of the term "biological diversity".

The term's contracted form *biodiversity* may have been coined by W.G. Rosen in 1985 while planning the 1986 *National Forum on Biological Diversity* organized by the National Research Council (NRC). It first appeared in a publication in 1988 when sociobiologist E. O. Wilson used it as the title of the proceedings of that forum.

Since this period the term has achieved widespread use among biologists, environmentalists, political leaders and concerned citizens.

A similar term in the United States is *"natural heritage."* It pre-dates the others and is more accepted by the wider audience interested in conservation. Broader than biodiversity, it includes geology and landforms.

Definitions

"Biodiversity" is most commonly used to replace the more clearly defined and long established terms, species diversity and species richness. Biologists most often define biodiversity as the "totality of genes, species and ecosystems of a region". An advantage of this definition is that it seems to describe most circumstances and presents a unified view of the traditional types of biological variety previously identified:

A sampling of fungi collected during summer 2008 in Northern Saskatchewan mixed woods, near LaRonge is an example regarding the species diversity of fungi. In this photo, there are also leaf lichens and mosses.

- taxonomic diversity (usually measured at the species diversity level)

- ecological diversity often viewed from the perspective of ecosystem diversity

- morphological diversity which stems from genetic diversity

- functional diversity which is a measure of the number of functionally disparate species within a population (e.g. different feeding mechanism, different motility, predator vs prey, etc.)

In 2003 Professor Anthony Campbell at Cardiff University, UK and the Darwin Centre, Pembrokeshire, defined a fourth level: Molecular Diversity.

This multilevel construct is consistent with Datman and Lovejoy. An explicit definition consistent with this interpretation was first given in a paper by Bruce A. Wilcox commissioned by the International Union for the Conservation of Nature and Natural Resources (IUCN) for the 1982 World National Parks Conference. Wilcox's definition was "Biological diversity is the variety of life forms...at all levels of biological systems (i.e., molecular, organismic, population, species and ecosystem)...". The 1992 United Nations Earth Summit defined "biological diversity" as "the variability among living organisms from all sources, including, 'inter alia', terrestrial, marine and other aquatic ecosystems and the ecological complexes of which they are part: this includes diversity within species, between species and of ecosystems". This definition is used in the United Nations Convention on Biological Diversity.

One textbook's definition is "variation of life at all levels of biological organization".

Genetically biodiversity can be defined as the diversity of alleles, genes and organisms. They study processes such as mutation and gene transfer that drive evolution.

Measuring diversity at one level in a group of organisms may not precisely correspond to diversity at other levels. However, tetrapod (terrestrial vertebrates) taxonomic and ecological diversity shows a very close correlation.

Distribution

Biodiversity is not evenly distributed, rather it varies greatly across the globe as well as within regions. Among other factors, the diversity of all living things (biota) depends on temperature, precipitation, altitude, soils, geography and the presence of other species. The study of the spatial distribution of organisms, species and ecosystems, is the science of biogeography.

A conifer forest in the Swiss Alps (National Park)

Diversity consistently measures higher in the tropics and in other localized regions such as the Cape Floristic Region and lower in polar regions generally. Rain forests that have had wet climates for a long time, such as Yasuní National Park in Ecuador, have particularly high biodiversity.

Terrestrial biodiversity is thought to be up to 25 times greater than ocean biodiversity. A recently discovered method put the total number of species on Earth at 8.7 million, of which 2.1 million were estimated to live in the ocean. However, this estimate seems to under-represent the diversity of microorganisms.

Latitudinal Gradients

Generally, there is an increase in biodiversity from the poles to the tropics. Thus localities at lower latitudes have more species than localities at higher latitudes. This is often referred to as the latitudinal gradient in species diversity. Several ecological mechanisms may contribute to the gradient, but the ultimate factor behind many of them is the greater mean temperature at the equator compared to that of the poles.

Even though terrestrial biodiversity declines from the equator to the poles, some studies claim that this characteristic is unverified in aquatic ecosystems, especially in marine ecosystems. The latitudinal distribution of parasites does not appear to follow this rule.

Hotspots

A biodiversity hotspot is a region with a high level of endemic species that is under threat from humans. The term hotspot was introduced in 1988 by Norman Myers. While hotspots are spread all over the world, the majority are forest areas and most are located in the tropics.

Brazil's Atlantic Forest is considered one such hotspot, containing roughly 20,000 plant species, 1,350 vertebrates and millions of insects, about half of which occur nowhere else. The island of Madagascar and India are also particularly notable. Colombia is characterized by high biodiversity, with the highest rate of species by area unit worldwide and it has the largest number of endemics (species that are not found naturally anywhere else) of any country. About 10% of the species of the Earth can be found in Colombia, including over 1,900 species of bird, more than in Europe and North America combined, Colombia has 10% of the world's mammals species, 14% of the amphibian species and 18% of the bird species of the world. Madagascar dry deciduous forests and lowland rainforests possess a high ratio of endemism. Since the island separated from mainland Africa 66 million years ago, many species and ecosystems have evolved independently. Indonesia's 17,000 islands cover 735,355 square miles (1,904,560 km²) and contain 10% of the world's flowering plants, 12% of mammals and 17% of reptiles, amphibians and birds—along with nearly 240 million people. Many regions of high biodiversity and/or endemism arise from specialized habitats which require unusual adaptations, for example, alpine environments in high mountains, or Northern European peat bogs.

Accurately measuring differences in biodiversity can be difficult. Selection bias amongst researchers may contribute to biased empirical research for modern estimates of biodiversity. In 1768, Rev. Gilbert White succinctly observed of his Selborne, Hampshire *"all nature is so full, that that district produces the most variety which is the most examined."*

Evolution and History

Biodiversity is the result of 3.5 billion years of evolution. The origin of life has not been definitely established by science, however some evidence suggests that life may already have been well-established only a few hundred million years after the formation of the Earth. Until approximately 600 million years ago, all life consisted of archaea, bacteria, protozoans and similar single-celled organisms.

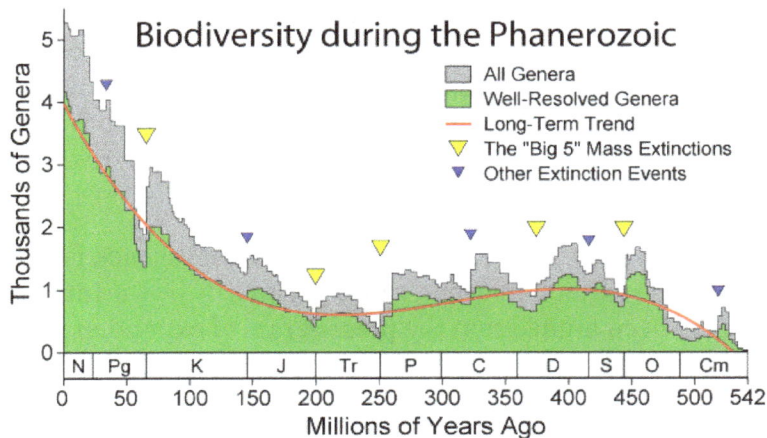

Apparent marine fossil diversity during the Phanerozoic

The history of biodiversity during the Phanerozoic (the last 540 million years), starts with rapid growth during the Cambrian explosion—a period during which nearly every phylum of multicellular organisms first appeared. Over the next 400 million years or so, invertebrate diversity showed little overall trend and vertebrate diversity shows an overall exponential trend. This dramatic rise in diversity was marked by periodic, massive losses of diversity classified as mass extinction events. A significant loss occurred when rainforests collapsed in the carboniferous. The worst was the Permian-Triassic extinction event, 251 million years ago. Vertebrates took 30 million years to recover from this event.

The fossil record suggests that the last few million years featured the greatest biodiversity in history. However, not all scientists support this view, since there is uncertainty as to how strongly the fossil record is biased by the greater availability and preservation of recent geologic sections. Some scientists believe that corrected for sampling artifacts, modern biodiversity may not be much different from biodiversity 300 million years ago., whereas others consider the fossil record reasonably reflective of the diversification of life. Estimates of the present global macroscopic species diversity vary from 2 million to 100 million, with a best estimate of somewhere near 9 million, the vast majority arthropods. Diversity appears to increase continually in the absence of natural selection.

Evolutionary Diversification

The existence of a "global carrying capacity", limiting the amount of life that can live at once, is debated, as is the question of whether such a limit would also cap the number of species. While records of life in the sea shows a logistic pattern of growth, life on land (insects, plants and tetrapods) shows an exponential rise in diversity. As one author states, "Tetrapods have not yet invaded 64 per cent of potentially habitable modes and it could be that without human influence the ecological and taxonomic diversity of tetrapods would continue to increase in an exponential fashion until most or all of the available ecospace is filled."

On the other hand, changes through the Phanerozoic correlate much better with the hyperbolic model (widely used in population biology, demography and macrosociology, as well as fossil biodiversity) than with exponential and logistic models. The latter models imply that changes in diversity are guided by a first-order positive feedback (more ancestors, more descendants) and/or a negative feedback arising from resource limitation. Hyperbolic model implies a second-order positive feedback. The hyperbolic pattern of the world population growth arises from a second-order positive feedback between the population size and the rate of technological growth. The hyperbolic character of biodiversity growth can be similarly accounted for by a feedback between diversity and community structure complexity. The similarity between the curves of biodiversity and human population probably comes from the fact that both are derived from the interference of the hyperbolic trend with cyclical and stochastic dynamics.

Most biologists agree however that the period since human emergence is part of a new mass extinction, named the Holocene extinction event, caused primarily by the impact humans are having on the environment. It has been argued that the present rate of extinction is sufficient to eliminate most species on the planet Earth within 100 years.

New species are regularly discovered (on average between 5–10,000 new species each year, most of them insects) and many, though discovered, are not yet classified (estimates are that nearly

90% of all arthropods are not yet classified). Most of the terrestrial diversity is found in tropical forests and in general, land has more species than the ocean; some 8.7 million species may exists on Earth, of which some 2.1 million live in the ocean

Biodiversity and Ecosystem Services

Summer field in Belgium (Hamois). The blue flowers are *Centaurea cyanus* and the red are *Papaver rhoeas*.

The Balance of Evidence

"Ecosystem services are the suite of benefits that ecosystems provide to humanity."

These services come in three flavors:

1. Provisioning services which involve the production of renewable resources (e.g.: food, wood, fresh water)

2. Regulating services which are those that lessen environmental change (e.g.: climate regulation, pest/disease control)

3. Cultural services represent human value and enjoyment (e.g.: landscape aesthetics, cultural heritage, outdoor recreation and spiritual significance)

There have been many claims about biodiversity's effect on these ecosystem services, especially provisioning and regulating services. After an exhaustive survey through peer-reviewed literature to evaluate 36 different claims about biodiversity's effect on ecosystem services, 14 of those claims have been validated, 6 demonstrate mixed support or are unsupported, 3 are incorrect and 13 lack enough evidence to draw definitive conclusions.

Services Enhanced by Biodiversity

Provisioning Services

- Greater species diversity of plants increases fodder yield (synthesis of 271 experimental studies).

- Greater genetic diversity of plants (i.e.: diversity within a single species) increases overall crop yield (synthesis of 575 experimental studies). Although another review of 100 experimental studies reports mixed evidence.

- Greater species diversity of trees increases overall wood production (Synthesis of 53 experimental studies). However, there is not enough data to draw a conclusion about the effect of tree trait diversity on wood production.

Regulating Services

- Greater species diversity of fish increases the stability of fisheries yield (Synthesis of 8 observational studies)

- Greater species diversity of natural pest enemies decreases herbivorous pest populations (Data from two separate reviews; Synthesis of 266 experimental and observational studies; Synthesis of 18 observational studies. Although another review of 38 experimental studies found mixed support for this claim, suggesting that in cases where mutual intraguild predation occurs, a single predatory species is often more effective

- Greater species diversity of plants decreases disease prevalence on plants (Synthesis of 107 experimental studies)

- Greater species diversity of plants increases resistance to plant invasion (Data from two separate reviews; Synthesis of 105 experimental studies; Synthesis of 15 experimental studies)

- Greater species diversity of plants increases carbon sequestration, but note that this finding only relates to actual uptake of carbon dioxide and not long term storage (Synthesis of 479 experimental studies)

- Greater species diversity of plants increases soil nutrient remineralization (Synthesis of 103 experimental studies)

- Greater species diversity of plants increases soil organic matter (Synthesis of 85 experimental studies)

Services with Mixed Evidence

Provisioning Services

- None to date

Regulating Services

- Greater species diversity of plants may or may not decrease herbivorous pest populations. Data from two separate reviews suggest that greater diversity decreases pest populations (Synthesis of 40 observational studies; Synthesis of 100 experimental studies). One review found mixed evidence (Synthesis of 287 experimental studies), while another found contrary evidence (Synthesis of 100 experimental studies)

- Greater species diversity of animals may or may not decrease disease prevalence on those animals (Synthesis of 45 experimental and observational studies), although a 2013 study offers more support showing that biodiversity may in fact enhance disease resistance within animal communities, at least in amphibian frog ponds. Many more studies must be published in support of diversity to sway the balance of evidence will be such that we can draw a general rule on this service.

- Greater species and trait diversity of plants may or may not increase long term carbon storage (Synthesis of 33 observational studies)

- Greater pollinator diversity may or may not increase pollination (Synthesis of 7 observational studies), but a publication from March 2013 suggests that increased native pollinator diversity enhances pollen deposition (although not necessarily fruit set as the authors would have you believe, for details explore their lengthy supplementary material).

Services for Which Biodiversity is A Hindrance

Provisioning services

- Greater species diversity of plants reduces primary production (Synthesis of 7 experimental studies)

Regulating Services

- Greater genetic and species diversity of a number of organisms reduces freshwater purification (Synthesis of 8 experimental studies, although an attempt by the authors to investigate the effect of detritivore diversity on freshwater purification was unsuccessful due to a lack of available evidence (only 1 observational study was found

Provisioning Services

- Effect of species diversity of plants on biofuel yield (In a survey of the literature, the investigators only found 3 studies)

- Effect of species diversity of fish on fishery yield (In a survey of the literature, the investigators only found 4 experimental studies and 1 observational study)

Regulating Services

- Effect of species diversity on the stability of biofuel yield (In a survey of the literature, the investigators did not find any studies)

- Effect of species diversity of plants on the stability of fodder yield (In a survey of the literature, the investigators only found 2 studies)

- Effect of species diversity of plants on the stability of crop yield (In a survey of the literature, the investigators only found 1 study)

- Effect of genetic diversity of plants on the stability of crop yield (In a survey of the literature, the investigators only found 2 studies)

- Effect of diversity on the stability of wood production (In a survey of the literature, the investigators could not find any studies)

- Effect of species diversity of multiple taxa on erosion control (In a survey of the literature, the investigators could not find any studies – they did however find studies on the effect of species diversity and root biomass)

- Effect of diversity on flood regulation (In a survey of the literature, the investigators could not find any studies)

- Effect of species and trait diversity of plants on soil moisture (In a survey of the literature, the investigators only found 2 studies)

Other sources have reported somewhat conflicting results and in 1997 Robert Costanza and colleagues reported the estimated global value of ecosystem services (not captured in traditional markets) at an average of $33 trillion annually.

Since the stone age, species loss has accelerated above the average basal rate, driven by human activity. Estimates of species losses are at a rate 100-10,000 times as fast as is typical in the fossil record. Biodiversity also affords many non-material benefits including spiritual and aesthetic values, knowledge systems and education.

Biodiversity and Agriculture

Agricultural diversity can be divided into two categories: intraspecific diversity, which includes the genetic variety within a single species, like the potato (*Solanum tuberosum*) that is composed of many different forms and types (e.g.: in the U.S. we might compare russet potatoes with new potatoes or purple potatoes, all different, but all part of the same species, *S. tuberosum*).

Amazon Rainforest in South America

The other category of agricultural diversity is called interspecific diversity and refers to the number and types of different species. Thinking about this diversity we might note that many small vegetable farmers grow many different crops like potatoes and also carrots, peppers, lettuce etc.

Agricultural diversity can also be divided by whether it is 'planned' diversity or 'associated' diversity. This is a functional classification that we impose and not an intrinsic feature of life or diversity. Planned diversity includes the crops which a farmer has encouraged, planted or raised (e.g.: crops, covers, symbionts and livestock, among others), which can be contrasted with the associated diversity that arrives among the crops, uninvited (e.g.: herbivores, weed species and pathogens, among others).

The control of associated biodiversity is one of the great agricultural challenges that farmers face. On monoculture farms, the approach is generally to eradicate associated diversity using a suite of biologically destructive pesticides, mechanized tools and transgenic engineering techniques, then to rotate crops. Although some polyculture farmers use the same techniques, they also employ integrated pest management strategies as well as strategies that are more labor-intensive, but generally less dependent on capital, biotechnology and energy.

Interspecific crop diversity is, in part, responsible for offering variety in what we eat. Intraspecific diversity, the variety of alleles within a single species, also offers us choice in our diets. If a crop fails in a monoculture, we rely on agricultural diversity to replant the land with something new. If a wheat crop is destroyed by a pest we may plant a hardier variety of wheat the next year, relying on intraspecific diversity. We may forgo wheat production in that area and plant a different species altogether, relying on interspecific diversity. Even an agricultural society which primarily grows monocultures, relies on biodiversity at some point.

- The Irish potato blight of 1846 was a major factor in the deaths of one million people and the emigration of about two million. It was the result of planting only two potato varieties, both vulnerable to the blight, *Phytophthora infestans*, which arrived in 1845

- When rice grassy stunt virus struck rice fields from Indonesia to India in the 1970s, 6,273 varieties were tested for resistance. Only one was resistant, an Indian variety and known to science only since 1966. This variety formed a hybrid with other varieties and is now widely grown.

- Coffee rust attacked coffee plantations in Sri Lanka, Brazil and Central America in 1970. A resistant variety was found in Ethiopia. The diseases are themselves a form of biodiversity.

Monoculture was a contributing factor to several agricultural disasters, including the European wine industry collapse in the late 19th century and the US southern corn leaf blight epidemic of 1970.

Although about 80 percent of humans' food supply comes from just 20 kinds of plants, humans use at least 40,000 species. Many people depend on these species for food, shelter and clothing. Earth's surviving biodiversity provides resources for increasing the range of food and other products suitable for human use, although the present extinction rate shrinks that potential.

Biodiversity and Human Health

Biodiversity's relevance to human health is becoming an international political issue, as scientific evidence builds on the global health implications of biodiversity loss. This issue is closely linked with the issue of climate change, as many of the anticipated health risks of climate change are associated with changes in biodiversity (e.g. changes in populations and distribution of disease vectors, scarcity of fresh water, impacts on agricultural biodiversity and food resources etc.) This is because the species most likely to disappear are those that buffer against infectious disease transmission, while surviving species tend to be the ones that increase disease transmission, such as that of West Nile Virus, Lyme disease and Hantavirus, according to a study done co-authored by Felicia Keesing, an ecologist at Bard College and Drew Harvell, associate director for Environment of the Atkinson Center for a Sustainable Future (ACSF) at Cornell University.

The diverse forest canopy on Barro Colorado Island, Panama, yielded this display of different fruit

The growing demand and lack of drinkable water on the planet presents an additional challenge to the future of human health. Partly, the problem lies in the success of water suppliers to increase supplies and failure of groups promoting preservation of water resources. While the distribution of clean water increases, in some parts of the world it remains unequal. According to *2008 World Population Data Sheet*, only 62% of least developed countries are able to access clean water.

Some of the health issues influenced by biodiversity include dietary health and nutrition security, infectious disease, medical science and medicinal resources, social and psychological health. Biodiversity is also known to have an important role in reducing disaster risk and in post-disaster relief and recovery efforts.

Biodiversity provides critical support for drug discovery and the availability of medicinal resources. A significant proportion of drugs are derived, directly or indirectly, from biological sources: at least 50% of the pharmaceutical compounds on the US market are derived from plants, animals and micro-organisms, while about 80% of the world population depends on medicines from nature (used in either modern or traditional medical practice) for primary healthcare. Only a tiny fraction of wild species has been investigated for medical potential. Biodiversity has been critical to advances throughout the field of bionics. Evidence from market analysis and biodiversity science indicates that the decline in output from the pharmaceutical sector since the mid-1980s can be attributed to a move away from natural product exploration ("bioprospecting") in favor of genomics and synthetic chemistry, indeed claims about the value of undiscovered pharmaceuticals may not provide enough incentive for companies in free markets to search for them because of the high cost of development; meanwhile, natural products have a long history of supporting significant economic and health innovation. Marine ecosystems are particularly important, although inappropriate bioprospecting can increase biodiversity loss, as well as violating the laws of the communities and states from which the resources are taken.

Biodiversity, Business and Industry

Many industrial materials derive directly from biological sources. These include building materials, fibers, dyes, rubber and oil. Biodiversity is also important to the security of resources such as water, timber, paper, fiber and food. As a result, biodiversity loss is a significant risk factor in business development and a threat to long term economic sustainability.

Agriculture production, pictured is a tractor and a chaser bin

Biodiversity, Leisure, Cultural and Aesthetic Value

Biodiversity enriches leisure activities such as hiking, birdwatching or natural history study. Biodiversity inspires musicians, painters, sculptors, writers and other artists. Many cultures view themselves as an integral part of the natural world which requires them to respect other living organisms.

Popular activities such as gardening, fishkeeping and specimen collecting strongly depend on biodiversity. The number of species involved in such pursuits is in the tens of thousands, though the majority do not enter commerce.

The relationships between the original natural areas of these often exotic animals and plants and commercial collectors, suppliers, breeders, propagators and those who promote their understanding and enjoyment are complex and poorly understood. The general public responds well to exposure to rare and unusual organisms, reflecting their inherent value.

Philosophically it could be argued that biodiversity has intrinsic aesthetic and spiritual value to mankind *in and of itself*. This idea can be used as a counterweight to the notion that tropical forests and other ecological realms are only worthy of conservation because of the services they provide.

Biodiversity and Ecological Services

Biodiversity supports many ecosystem services:

"There is now unequivocal evidence that biodiversity loss reduces the efficiency by which ecological communities capture biologically essential resources, produce biomass, decompose and recycle biologically essential nutrients... There is mounting evidence that biodiversity increases the stability of ecosystem functions through time... Diverse communities are more productive because they contain key species that have a large influence on productivity and differences in functional traits among organisms increase total resource capture... The impacts of diversity loss on ecological processes might be sufficiently large to rival the impacts of many other global drivers of environmental change... Maintaining multiple ecosystem processes at multiple places and times requires higher levels of biodiversity than does a single process at a single place and time."

Eagle Creek, Oregon hiking

It plays a part in regulating the chemistry of our atmosphere and water supply. Biodiversity is directly involved in water purification, recycling nutrients and providing fertile soils. Experiments with controlled environments have shown that humans cannot easily build ecosystems to support human needs; for example insect pollination cannot be mimicked, and that activity alone represented between $2.1-14.6 billions in 2003.

Number of Species

Species	Earth			Ocean		
	Catalogued	Predicted	±SE	Catalogued	Predicted	±SE
Eukaryotes						
Animalia	953,434	7,770,000	958,000	171,082	2,150,000	145,000
Chromista	13,033	27,500	30,500	4,859	7,400	9,640
Fungi	43,271	611,000	297,000	1,097	5,320	11,100
Plantae	215,644	298,000	8,200	8,600	16,600	9,130
Protozoa	8,118	36,400	6,690	8,118	36,400	6,690
Total	1,233,500	8,740,000	1,300,000	193,756	2,210,000	182,000
Prokaryotes						
Archaea	502	455	160	1	1	0
Bacteria	10,358	9,680	3,470	652	1,320	436
Total	10,860	10,100	3,630	653	1,320	436
Grand Total	**1,244,360**	**8,750,000**	**1,300,000**	**194,409**	**2,210,000**	**182,000**

Predictions for prokaryotes represent a lower bound because they do not consider undescribed higher taxa. For protozoa, the ocean database was substantially more complete than the database for the entire Earth so we only used the former to estimate the total number of species in this taxon. All predictions were rounded to three significant digits.
doi:10.1371/journal.pbio.1001127.t002

Discovered and predicted total number of species on land and in the oceans

According to Mora and colleagues, the total number of terrestrial species is estimated to be around 8.7 million while the number of oceanic species is much lower, estimated at 2.2 million. The authors note that these estimates are strongest for eukaryotic organisms and likely represent the lower bound of prokaryote diversity. Other estimates include:

- 220,000 vascular plants, estimated using the species-area relation method

- 0.7-1 million marine species

- 10–30 million insects; (of some 0.9 million we know today)

- 5–10 million bacteria;

- 1.5-3 million fungi, estimates based on data from the tropics, long-term non-tropical sites and molecular studies that have revealed cryptic speciation. Some 0.075 million species of fungi had been documented by 2001)

- 1 million mites

- The number of microbial species is not reliably known, but the Global Ocean Sampling Expedition dramatically increased the estimates of genetic diversity by identifying an enormous number of new genes from near-surface plankton samples at various marine locations, initially over the 2004-2006 period. The findings may eventually cause a significant change in the way science defines species and other taxonomic categories.

Since the rate of extinction has increased, many extant species may become extinct before they are described.

Species Loss Rates

No longer do we have to justify the existence of humid tropical forests on the feeble grounds that they might carry plants with drugs that cure human disease. *Gaia theory* forces us to see that they offer much more than this. Through their capacity to evapotranspirate vast volumes of water vapor, they serve to keep the planet cool by wearing a sunshade of white reflecting cloud. Their replacement by cropland could precipitate a disaster that is global in scale.

During the last century, decreases in biodiversity have been increasingly observed. In 2007, German Federal Environment Minister Sigmar Gabriel cited estimates that up to 30% of all species will be extinct by 2050. Of these, about one eighth of known plant species are threatened with extinction. Estimates reach as high as 140,000 species per year (based on Species-area theory). This figure indicates unsustainable ecological practices, because few species emerge each year. Almost all scientists acknowledge that the rate of species loss is greater now than at any time in human history, with extinctions occurring at rates hundreds of times higher than background extinction rates. As of 2012, some studies suggest that 25% of all mammal species could be extinct in 20 years.

In absolute terms, the planet has lost 52% of its biodiversity since 1970 according to a 2014 study by the World Wildlife Fund. The Living Planet Report 2014 claims that "the number of mammals, birds, reptiles, amphibians and fish across the globe is, on average, about half the size it was 40 years ago". Of that number, 39% accounts for the terrestrial wildlife gone, 39% for the marine wildlife gone and 76% for the freshwater wildlife gone. Biodiversity took the biggest hit in Latin America, plummeting 83 percent. High-income countries showed a 10% increase in biodiversity, which was canceled out by a loss in low-income countries. This is despite the fact that high-income countries use five times the ecological resources of low-income countries, which was explained as a result of process whereby wealthy nations are outsourcing resource depletion to poorer nations, which are suffering the greatest ecosystem losses.

Threats

In 2006 many species were formally classified as rare or endangered or threatened; moreover, scientists have estimated that millions more species are at risk which have not been formally recognized. About 40 percent of the 40,177 species assessed using the IUCN Red List criteria are now listed as threatened with extinction—a total of 16,119.

Jared Diamond describes an "Evil Quartet" of habitat destruction, overkill, introduced species and secondary extinctions. Edward O. Wilson prefers the acronym HIPPO, standing for Habitat destruction, Invasive species, Pollution, human over-Population and Over-harvesting. The most authoritative classification in use today is IUCN's Classification of Direct Threats which has been adopted by major international conservation organizations such as the US Nature Conservancy, the World Wildlife Fund, Conservation International and BirdLife International.

Habitat Destruction

Habitat destruction has played a key role in extinctions, especially related to tropical forest destruction. Factors contributing to habitat loss are: overconsumption, overpopulation, land use change, deforestation, pollution (air pollution, water pollution, soil contamination) and global warming or climate change.

Deforestation and increased road-building in the Amazon Rainforest are a significant concern because of increased human encroachment upon wild areas, increased resource extraction and further threats to biodiversity.

Habitat size and numbers of species are systematically related. Physically larger species and those living at lower latitudes or in forests or oceans are more sensitive to reduction in habitat area. Conversion to "trivial" standardized ecosystems (e.g., monoculture following deforestation) effectively destroys habitat for the more diverse species that preceded the conversion. In some countries lack of property rights or lax law/regulatory enforcement necessarily leads to biodiversity loss (degradation costs having to be supported by the community).

A 2007 study conducted by the National Science Foundation found that biodiversity and genetic diversity are codependent—that diversity among species requires diversity within a species and vice versa. "If any one type is removed from the system, the cycle can break down and the community becomes dominated by a single species." At present, the most threatened ecosystems are found in fresh water, according to the Millennium Ecosystem Assessment 2005, which was confirmed by the "Freshwater Animal Diversity Assessment", organised by the biodiversity platform and the French *Institut de recherche pour le développement* (MNHNP).

Co-extinctions are a form of habitat destruction. Co-extinction occurs when the extinction or decline in one accompanies the other, such as in plants and beetles.

Introduced and Invasive Species

Barriers such as large rivers, seas, oceans, mountains and deserts encourage diversity by enabling independent evolution on either side of the barrier, via the process of allopatric speciation. The term invasive species is applied to species that breach the natural barriers that would normally keep them constrained. Without barriers, such species occupy new territory, often supplanting native species by occupying their niches, or by using resources that would normally sustain native species.

Male *Lophura nycthemera* (silver pheasant), a native of East Asia that has been introduced into parts of Europe for ornamental reasons

The number of species invasions has been on the rise at least since the beginning of the 1900s. Species are increasingly being moved by humans (on purpose and accidentally). In some cases the invaders are causing drastic changes and damage to their new habitats (e.g.: zebra mussels and the emerald ash borer in the Great Lakes region and the lion fish along the North American Atlantic coast). Some evidence suggests that invasive species are competitive in their new habitats because they are subject to less pathogen disturbance. Others report confounding evidence that occasionally suggest that species-rich communities harbor many native and exotic species simultaneously while some say that diverse ecosystems are more resilient and resist invasive plants and animals. An important question is, "do invasive species cause extinctions?" Many studies cite effects of invasive species on natives, but not extinctions. Invasive species seem to increase local (i.e.: alpha diversity) diversity, which decreases turnover of diversity (i.e.: beta diversity). Overall gamma diversity may be lowered because species are going extinct because of other causes, but even some of the most insidious invaders (e.g.: Dutch elm disease, emerald ash borer, chestnut blight in North America) have not caused their host species to become extinct. Extirpation, population decline and homogenization of regional biodiversity are much more common. Human activities have frequently been the cause of invasive species circumventing their barriers, by introducing them for food and other purposes. Human activities therefore allow species to migrate to new areas (and thus become invasive) occurred on time scales much shorter than historically have been required for a species to extend its range.

Not all introduced species are invasive, nor all invasive species deliberately introduced. In cases such as the zebra mussel, invasion of US waterways was unintentional. In other cases, such as mongooses in Hawaii, the introduction is deliberate but ineffective (nocturnal rats were not vulnerable to the diurnal mongoose). In other cases, such as oil palms in Indonesia and Malaysia, the introduction produces substantial economic benefits, but the benefits are accompanied by costly unintended consequences.

Finally, an introduced species may unintentionally injure a species that depends on the species it replaces. In Belgium, Prunus spinosa from Eastern Europe leafs much sooner than its West European counterparts, disrupting the feeding habits of the *Thecla betulae* butterfly (which feeds on the leaves). Introducing new species often leaves endemic and other local species unable to compete with the exotic species and unable to survive. The exotic organisms may be predators, parasites, or may simply outcompete indigenous species for nutrients, water and light.

At present, several countries have already imported so many exotic species, particularly agricultural and ornamental plants, that their own indigenous fauna/flora may be outnumbered. For example, the introduction of kudzu from Southeast Asia to Canada and the United States has threatened biodiversity in certain areas.

Genetic Pollution

Endemic species can be threatened with extinction through the process of genetic pollution, i.e. uncontrolled hybridization, introgression and genetic swamping. Genetic pollution leads to homogenization or replacement of local genomes as a result of either a numerical and/or fitness advantage of an introduced species. Hybridization and introgression are side-effects of introduction and invasion. These phenomena can be especially detrimental to rare species that come into contact with more abundant ones. The abundant species can interbreed with the rare species, swamping its gene pool. This problem is not always apparent from morphological (outward appearance) observations alone. Some degree of gene flow is normal adaptation and not all gene and genotype constellations can be preserved. However, hybridization with or without introgression may, nevertheless, threaten a rare species' existence.

Overexploitation

Overexploitation occurs when a resource is consumed at an unsustainable rate. This occurs on land in the form of overhunting, excessive logging, poor soil conservation in agriculture and the illegal wildlife trade.

About 25% of world fisheries are now overfished to the point where their current biomass is less than the level that maximizes their sustainable yield.

The overkill hypothesis, a pattern of large animal extinctions connected with human migration patterns, can be used explain why megafaunal extinctions can occur within a relatively short time period.

Hybridization, Genetic Pollution/Erosion and Food Security

In agriculture and animal husbandry, the Green Revolution popularized the use of conventional hybridization to increase yield. Often hybridized breeds originated in developed countries and were further hybridized with local varieties in the developing world to create high yield strains resistant to local climate and diseases. Local governments and industry have been pushing hybridization. Formerly huge gene pools of various wild and indigenous breeds have collapsed causing widespread genetic erosion and genetic pollution. This has resulted in loss of genetic diversity and biodiversity as a whole.

The Yecoro wheat (right) cultivar is sensitive to salinity, plants resulting from a hybrid cross with cultivar W4910 (left) show greater tolerance to high salinity

(GM organisms) have genetic material altered by genetic engineering procedures such as recombinant DNA technology. GM crops have become a common source for genetic pollution, not only of wild varieties but also of domesticated varieties derived from classical hybridization.

Genetic erosion coupled with genetic pollution may be destroying unique genotypes, thereby creating a hidden crisis which could result in a severe threat to our food security. Diverse genetic material could cease to exist which would impact our ability to further hybridize food crops and livestock against more resistant diseases and climatic changes.

Climate Change

Global warming is also considered to be a major potential threat to global biodiversity in the future. For example, coral reefs - which are biodiversity hotspots - will be lost within the century if global warming continues at the current trend.

Polar bears on the sea ice of the Arctic Ocean, near the North Pole. Climate change has started affecting bear populations.

Climate change has seen many claims about potential to affect biodiversity but evidence supporting the statement is tenuous. Increasing atmospheric carbon dioxide certainly affects plant morphology and is acidifying oceans, and temperature affects species ranges, phenology, and weather, but the major impacts that have been predicted are still just *potential* impacts. We have not documented major extinctions yet, even as climate change drastically alters the biology of many species.

In 2004, an international collaborative study on four continents estimated that 10 percent of species would become extinct by 2050 because of global warming. "We need to limit climate change or we wind up with a lot of species in trouble, possibly extinct," said Dr. Lee Hannah, a co-author of the paper and chief climate change biologist at the Center for Applied Biodiversity Science at Conservation International.

A recent study predicts that up to 35% of the world terrestrial carnivores and ungulates will be at higher risk of extinction by 2050 because of the joint effects of predicted climate and land-use change under business-as-usual human development scenarios.

Human Overpopulation

From 1950 to 2011, world population increased from 2.5 billion to 7 billion and is forecast to reach a plateau of more than 9 billion during the 21st century. Sir David King, former chief scientific adviser to the UK government, told a parliamentary inquiry: *"It is self-evident that the massive growth in the human population through the 20th century has had more impact on biodiversity than any other single factor."* At least until the middle of the 21st century, worldwide losses of pristine biodiverse land will probably depend much on the worldwide human birth rate.

According to a 2014 study by the World Wildlife Fund, the global human population already exceeds planet's biocapacity - it would take the equivalent of 1.5 Earths of biocapacity to meet our current demands. The report further points that if everyone on the planet had the Footprint of the average resident of Qatar, we would need 4.8 Earths and if we lived the lifestyle of a typical resident of the USA, we would need 3.9 Earths.

The Holocene Extinction

Rates of decline in biodiversity in this sixth mass extinction match or exceed rates of loss in the five previous mass extinction events in the fossil record. Loss of biodiversity results in the loss of natural capital that supplies ecosystem goods and services. From the perspective of the method known as Natural Economy the economic value of 17 ecosystem services for Earth's biosphere (calculated in 1997) has an estimated value of US$33 trillion ($3.3 \times 10^{13}$) per year.

Conservation

Conservation biology matured in the mid-20th century as ecologists, naturalists and other scientists began to research and address issues pertaining to global biodiversity declines.

The conservation ethic advocates management of natural resources for the purpose of sustaining biodiversity in species, ecosystems, the evolutionary process and human culture and society.

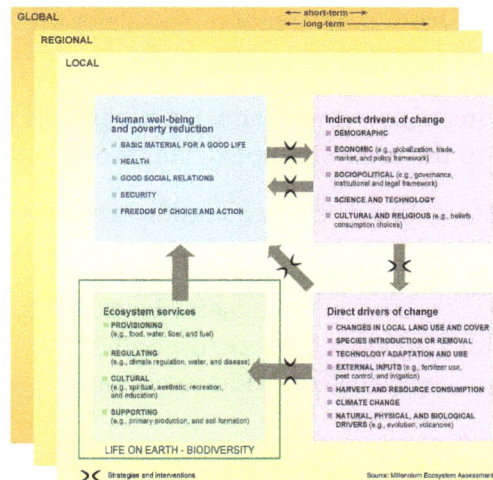

A schematic image illustrating the relationship between biodiversity, ecosystem services, human well-being and poverty. The illustration shows where conservation action, strategies and plans can influence the drivers of the current biodiversity crisis at local, regional, to global scales.

Conservation biology is reforming around strategic plans to protect biodiversity. Preserving global biodiversity is a priority in strategic conservation plans that are designed to engage public policy and concerns affecting local, regional and global scales of communities, ecosystems and cultures. Action plans identify ways of sustaining human well-being, employing natural capital, market capital and ecosystem services.

The retreat of Aletsch Glacier in the Swiss Alps (situation in 1979, 1991 and 2002), due to global warming.

In the EU Directive 1999/22/EC zoos are described as having a role in the preservation of the biodiversity of wildlife animals by conducting research or participation in breeding programs.

Protection and Restoration Techniques

Removal of exotic species will allow the species that they have negatively impacted to recover their ecological niches. Exotic species that have become pests can be identified taxonomically (e.g., with Digital Automated Identification SYstem (DAISY), using the barcode of life). Removal is practical only given large groups of individuals due to the economic cost.

As sustainable populations of the remaining native species in an area become assured, "missing" species that are candidates for reintroduction can be identified using databases such as the *Encyclopedia of Life* and the Global Biodiversity Information Facility.

- Biodiversity banking places a monetary value on biodiversity. One example is the Australian Native Vegetation Management Framework.

- Gene banks are collections of specimens and genetic material. Some banks intend to reintroduce banked species to the ecosystem (e.g., via tree nurseries).

- Reduction of and better targeting of pesticides allows more species to survive in agricultural and urbanized areas.

- Location-specific approaches may be less useful for protecting migratory species. One approach is to create wildlife corridors that correspond to the animals' movements. National and other boundaries can complicate corridor creation.

Protected Areas

Protected areas is meant for affording protection to wild animals and their habitat which also includes forest reserves and biosphere reserves. Protected areas have been set up all over the world with the specific aim of protecting and conserving plants and animals.

National Parks

National park and nature reserve is the area selected by governments or private organizations for special protection against damage or degradation with the objective of biodiversity and landscape conservation. National parks are usually owned and managed by national or state governments. A limit is placed on the number of visitors permitted to enter certain fragile areas. Designated trails or roads are created. The visitors are allowed to enter only for study, cultural and recreation purposes. Forestry operations, grazing of animals and hunting of animals are regulated. Exploitation of habitat or wildlife is banned.

Wildlife Sanctuary

Wildlife sanctuary aims only at conservation of species and have the following features:

1. The boundaries of the sanctuaries are not limited by state legislation.

2. The killing, hunting or capturing of any species is prohibited except by or under the control of the highest authority in the department which is responsible for the management of the sanctuary.

3. Private ownership may be allowed.

4. Forestry and other usages can also be permitted.

Forest Reserves

The forests play a vital role in harbouring more than 45,000 floral and 81,000 faunal species of which 5150 floral and 1837 faunal species are endemic. Plant and animal species confined to a specific geographical area are called endemic species. In reserved forests, rights to activities like hunting and grazing are sometimes given to communities living on the fringes of the forest, who sustain their livelihood partially or wholly from forest resources or products. The unclassed forests covers 6.4 percent of the total forest area and they are marked by the following characteristics:

1. They are large inaccessible forests.

2. Many of these are unoccupied.

3. They are ecologically and economically less important.

Steps to Conserve the Forest Cover

1. An extensive reforestation/afforestation program should be followed.

2. Alternative environment-friendly sources of fuel energy such as biogas other than wood should be used.

3. Loss of biodiversity due to forest fire is a major problem, immediate steps to prevent forest fire need to be taken.

4. Overgrazing by cattle can damage a forest seriously. Therefore, certain steps should be taken to prevent overgrazing by cattle.

5. Hunting and poaching should be banned.

Zoological Parks

In zoological parks or zoos, live animals are kept for public recreation, education and conservation purposes. Modern zoos offer veterinary facilities, provide opportunities for threatened species to breed in captivity and usually build environments that simulate the native habitats of the animals in their care. Zoos play a major role in creating awareness among common people about the need to conserve nature.

Botanical Gardens

Botanical garden is a garden in which plants are grown and displayed primarily for scientific and educational purposes. It consists of a collection of living plants, grown outdoors or under glass in greenhouses and conservatories. In addition, it includes a collection of dried plants or herbarium and such facilities as lecture rooms, laboratories, libraries, museums and experimental or research plantings.

Resource Allocation

Focusing on limited areas of higher potential biodiversity promises greater immediate return on investment than spreading resources evenly or focusing on areas of little diversity but greater interest in biodiversity.

A second strategy focuses on areas that retain most of their original diversity, which typically require little or no restoration. These are typically non-urbanized, non-agricultural areas. Tropical areas often fit both criteria, given their natively high diversity and relative lack of development.

Legal Status

A great deal of work is occurring to preserve the natural characteristics of Hopetoun Falls, Australia while continuing to allow visitor access.

International

- United Nations Convention on Biological Diversity (1992) and Cartagena Protocol on Biosafety;

- Convention on International Trade in Endangered Species (CITES);

- Ramsar Convention (Wetlands);

- Bonn Convention on Migratory Species;

- World Heritage Convention (indirectly by protecting biodiversity habitats)

- Regional Conventions such as the Apia Convention

- Bilateral agreements such as the Japan-Australia Migratory Bird Agreement.

Global agreements such as the Convention on Biological Diversity, give "sovereign national rights over biological resources" (not property). The agreements commit countries to "conserve biodiversity", "develop resources for sustainability" and "share the benefits" resulting from their use. Biodiverse countries that allow bioprospecting or collection of natural products, expect a share of the benefits rather than allowing the individual or institution that discovers/exploits the resource to capture them privately. Bioprospecting can become a type of biopiracy when such principles are not respected.

Sovereignty principles can rely upon what is better known as Access and Benefit Sharing Agreements (ABAs). The Convention on Biodiversity implies informed consent between the source country and the collector, to establish which resource will be used and for what and to settle on a fair agreement on benefit sharing.

National Level Laws

Biodiversity is taken into account in some political and judicial decisions:

- The relationship between law and ecosystems is very ancient and has consequences for biodiversity. It is related to private and public property rights. It can define protection for threatened ecosystems, but also some rights and duties (for example, fishing and hunting rights).

- Law regarding species is more recent. It defines species that must be protected because they may be threatened by extinction. The U.S. Endangered Species Act is an example of an attempt to address the "law and species" issue.

- Laws regarding gene pools are only about a century old. Domestication and plant breeding methods are not new, but advances in genetic engineering have led to tighter laws covering distribution of genetically modified organisms, gene patents and process patents. Governments struggle to decide whether to focus on for example, genes, genomes, or organisms and species.

Uniform approval for use of biodiversity as a legal standard has not been achieved, however. Bosselman argues that biodiversity should not be used as a legal standard, claiming that the remaining areas of scientific uncertainty cause unacceptable administrative waste and increase litigation without promoting preservation goals.

India passed the Biological Diversity Act in 2002 for the conservation of biological diversity in India. The Act also provides mechanisms for equitable sharing of benefits from the use of traditional biological resources and knowledge.

Analytical Limits

Taxonomic and Size Relationships

Less than 1% of all species that have been described have been studied beyond simply noting their existence. The vast majority of Earth's species are microbial. Contemporary biodiversity physics is "firmly fixated on the visible [macroscopic] world". For example, microbial life is metabolically and environmentally more diverse than multicellular life. "On the tree of life, based on analyses of small-subunit ribosomal RNA, visible life consists of barely noticeable twigs. The inverse relationship of size and population recurs higher on the evolutionary ladder—"to a first approximation, all multicellular species on Earth are insects". Insect extinction rates are high—supporting the Holocene extinction hypothesis.

References

- Kunin, W.E.; Gaston, Kevin, eds. (31 December 1996). The Biology of Rarity: Causes and consequences of rare—common differences. ISBN 978-0412633805. Retrieved 26 May 2015.

- Stearns, Beverly Peterson; Stearns, S. C.; Stearns, Stephen C. (2000). Watching, from the Edge of Extinction. Yale University Press. p. 1921. ISBN 978-0-300-08469-6. Retrieved 2014-12-27.

- Cockell, Charles; Koeberl, Christian; Gilmour, Iain (18 May 2006). Biological Processes Associated with Impact Events (1 ed.). Springer Science & Business Media. pp. 197–219. ISBN 978-3-540-25736-3.

- G. Miller; Scott Spoolman (2012). Environmental Science - Biodiversity Is a Crucial Part of the Earth's Natural Capital. Cengage Learning. p. 62. ISBN 1-133-70787-4. Retrieved 2014-12-27.

- Hamilton Raven, Peter; Brooks Johnson, George (2002). Biology. McGraw-Hill Education. p. 68. ISBN 978-0-07-112261-0. Retrieved 2013-07-07.

- Global Biodiversity Assessment: Summary for Policy-makers. Cambridge University Press. 1995. ISBN 978-0-521-56481-6. Annex 6, Glossary. Used as source by "Biodiversity", Glossary of terms related to the CBD, Belgian Clearing-House Mechanism. Retrieved 2006-04-26.

- Tor-Björn Larsson (2001). Biodiversity evaluation tools for European forests. Wiley-Blackwell. p. 178. ISBN 978-87-16-16434-6. Retrieved 28 June 2011.

- Morand, Serge; Krasnov, Boris R. (1 September 2010). The Biogeography of Host-Parasite Interactions. Oxford University Press. pp. 93–94. ISBN 978-0-19-956135-3. Retrieved 28 June 2011.

- McKee, Jeffrey K. (December 2004). Sparing Nature: The Conflict Between Human Population Growth and Earth's Biodiversity. Rutgers University Press. p. 108. ISBN 978-0-8135-3558-6. Retrieved 28 June 2011.

- Futuyma, Douglas J.; Shaffer, H. Bradley; Simberloff, Daniel, eds. (1 January 2009). Annual Reviews of Ecology, Evolution and Systematics: Vol 40 2009. Palo Alto, Calif.: Annual Reviews. pp. 573–592. ISBN 978-0-8243-1440-8.

- Hassan, Rashid M.; et al. (2006). Ecosystems and human well-being: current state and trends : findings of the Condition and Trends Working Group of the Millennium Ecosystem Assessment. Island Press. p. 105. ISBN 978-1-55963-228-7.

- Chen, Jim (2003). "Across the Apocalypse on Horseback: Imperfect Legal Responses to Biodiversity Loss". The Jurisdynamics of Environmental Protection: Change and the Pragmatic Voice in Environmental Law. Environmental Law Institute. p. 197. ISBN 978-1-58576-071-8.

- Ehrlich, Paul R.; Ehrlich, Anne H. (1983). Extinction: The Causes and Consequences of the Disappearance of Species. Ballantine Books. ISBN 978-0-345-33094-9.

- Borenstein, Seth (19 October 2015). "Hints of life on what was thought to be desolate early Earth". Excite. Yonkers, NY: Mindspark Interactive Network. Associated Press. Retrieved 2015-10-20.

- Visconti, Piero; et. al (February 2015). "Projecting global biodiversity indicators under future development scenarios". Conservation Letters. Wiley. doi:10.1111/conl.12159. Retrieved 2015-03-25.

- "Beantwoording vragen over fokken en doden van gezonde dieren in dierentuinen" (PDF) (in Dutch). Ministry of Economic Affairs (Netherlands). 25 March 2014. Retrieved 9 June 2014.

2

Esssential Concepts in Biodiversity

The chapter strategically encompasses and incorporates the major components and key concepts of biodiversity and provides a complete understanding of them. The topics discussed in this chapter are alpha diversity, extinction, defaunation, extinction event, snowball Earth, endemism and biodiversity hotspot. This chapter is a compilation of the various branches of biodiversity that form an integral part of the broader subject matter.

Ecosystem

An ecosystem is a community of living organisms in conjunction with the nonliving components of their environment (things like air, water and mineral soil), interacting as a system. These biotic and abiotic components are regarded as linked together through nutrient cycles and energy flows. As ecosystems are defined by the network of interactions among organisms, and between organisms and their environment, they can be of any size but usually encompass specific, limited spaces (although some scientists say that the entire planet is an ecosystem).

Coral reefs are a highly productive marine ecosystem.

Energy, water, nitrogen and soil minerals are other essential abiotic components of an ecosystem. The energy that flows through ecosystems is obtained primarily from the sun. It generally enters the system through photosynthesis, a process that also captures carbon from the atmosphere. By feeding on plants and on one another, animals play an important role in the movement of matter and energy through the system. They also influence the quantity of plant and microbial biomass present. By breaking down dead organic matter, decomposers release carbon back to the atmosphere and facilitate nutrient cycling by converting nutrients stored in dead biomass back to a form that can be readily used by plants and other microbes.

Rainforest ecosystems are rich in biodiversity. This is the Gambia River in Senegal's Niokolo-Koba National Park.

Ecosystems are controlled both by external and internal factors. External factors such as climate, the parent material that forms the soil, and topography control the overall structure of an ecosystem and the way things work within it, but are not themselves influenced by the ecosystem. Other external factors include time and potential biota. Ecosystems are dynamic entities—invariably, they are subject to periodic disturbances and are in the process of recovering from some past disturbance. Ecosystems in similar environments that are located in different parts of the world can have very different characteristics simply because they contain different species. The introduction of non-native species can cause substantial shifts in ecosystem function. Internal factors not only control ecosystem processes but are also controlled by them and are often subject to feedback loops. While the resource inputs are generally controlled by external processes like climate and parent material, the availability of these resources within the ecosystem is controlled by internal factors like decomposition, root competition or shading. Other internal factors include disturbance, succession and the types of species present. Although humans exist and operate within ecosystems, their cumulative effects are large enough to influence external factors like climate.

Biodiversity affects ecosystem function, as do the processes of disturbance and succession. Ecosystems provide a variety of goods and services upon which people depend; the principles of ecosystem management suggest that rather than managing individual species, natural resources should be managed at the level of the ecosystem itself. Classifying ecosystems into ecologically homogeneous units is an important step towards effective ecosystem management, but there is no single, agreed-upon way to do this.

History and Development

The term "ecosystem" was first used in a publication by British ecologist Arthur Tansley. Tansley devised the concept to draw attention to the importance of transfers of materials between organisms and their environment. He later refined the term, describing it as "The whole system, ... including not only the organism-complex, but also the whole complex of physical factors forming what we call the environment". Tansley regarded ecosystems not simply as natural units, but as mental isolates. Tansley later defined the spatial extent of ecosystems using the term ecotope.

G. Evelyn Hutchinson, a pioneering limnologist who was a contemporary of Tansley's, combined Charles Elton's ideas about trophic ecology with those of Russian geochemist Vladimir Vernadsky to suggest that mineral nutrient availability in a lake limited algal production which would, in turn, limit the abundance of animals that feed on algae. Raymond Lindeman took these ideas one step further to suggest that the flow of energy through a lake was the primary driver of the ecosystem. Hutchinson's students, brothers Howard T. Odum and Eugene P. Odum, further developed a "systems approach" to the study of ecosystems, allowing them to study the flow of energy and material through ecological systems.

Ecosystem Processes

Energy and carbon enter ecosystems through photosynthesis, are incorporated into living tissue, transferred to other organisms that feed on the living and dead plant matter, and eventually released through respiration. Most mineral nutrients, on the other hand, are recycled within ecosystems.

Ecosystems are controlled both by external and internal factors. External factors, also called state factors, control the overall structure of an ecosystem and the way things work within it, but are not themselves influenced by the ecosystem. The most important of these is climate. Climate determines the biome in which the ecosystem is embedded. Rainfall patterns and temperature seasonality determine the amount of water available to the ecosystem and the supply of energy available (by influencing photosynthesis). Parent material, the underlying geological material that gives rise to soils, determines the nature of the soils present, and influences the supply of mineral nutrients. Topography also controls ecosystem processes by affecting things like microclimate, soil development and the movement of water through a system. This may be the difference between the ecosystem present in wetland situated in a small depression on the landscape, and one present on an adjacent steep hillside.

Other external factors that play an important role in ecosystem functioning include time and potential biota. Ecosystems are dynamic entities—invariably, they are subject to periodic disturbances and are in the process of recovering from some past disturbance. Time plays a role in the development of soil from bare rock and the recovery of a community from disturbance. Similarly, the set of organisms that can potentially be present in an area can also have a major impact on ecosystems. Ecosystems in similar environments that are located in different parts of the world can end up doing things very differently simply because they have different pools of species present. The introduction of non-native species can cause substantial shifts in ecosystem function.

Unlike external factors, internal factors in ecosystems not only control ecosystem processes, but are also controlled by them. Consequently, they are often subject to feedback loops. While the resource inputs are generally controlled by external processes like climate and parent material, the availability of these resources within the ecosystem is controlled by internal factors like decomposition, root competition or shading. Other factors like disturbance, succession or the types of species present are also internal factors. Human activities are important in almost all ecosystems. Although humans exist and operate within ecosystems, their cumulative effects are large enough to influence external factors like climate.

Primary Production

Primary production is the production of organic matter from inorganic carbon sources. Over-whelmingly, this occurs through photosynthesis. The energy incorporated through this process supports life on earth, while the carbon makes up much of the organic matter in living and dead biomass, soil carbon and fossil fuels. It also drives the carbon cycle, which influences global climate via the greenhouse effect.

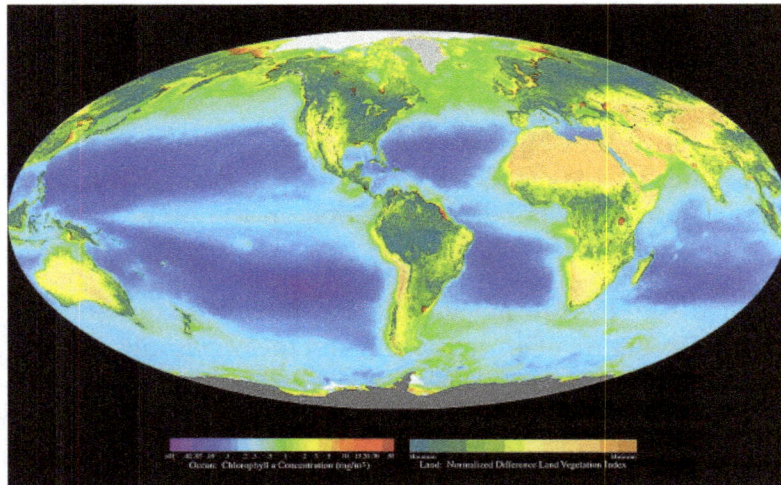

Global oceanic and terrestrial phototroph abundance, from September 1997 to August 2000. As an estimate of autotroph biomass, it is only a rough indicator of primary production potential, and not an actual estimate of it. Provided by the SeaWiFS Project, NASA/Goddard Space Flight Center and ORBIMAGE.

Through the process of photosynthesis, plants capture energy from light and use it to combine carbon dioxide and water to produce carbohydrates and oxygen. The photosynthesis carried out by all the plants in an ecosystem is called the gross primary production (GPP). About 48–60% of the GPP is consumed in plant respiration. The remainder, that portion of GPP that is not used up by respiration, is known as the net primary production (NPP). Total photosynthesis is limited by a range of environmental factors. These include the amount of light available, the amount of leaf area a plant has to capture light (shading by other plants is a major limitation of photosynthesis), rate at which carbon dioxide can be supplied to the chloroplasts to support photosynthesis, the availability of water, and the availability of suitable temperatures for carrying out photosynthesis.

Energy Flow

The carbon and energy incorporated into plant tissues (net primary production) is either consumed by animals while the plant is alive, or it remains uneaten when the plant tissue dies and becomes detritus. In terrestrial ecosystems, roughly 90% of the NPP ends up being broken down by decomposers. The remainder is either consumed by animals while still alive and enters the plant-based trophic system, or it is consumed after it has died, and enters the detritus-based trophic system. In aquatic systems, the proportion of plant biomass that gets consumed by herbivores is much higher. In trophic systems photosynthetic organisms are the primary producers. The organisms that consume their tissues are called primary consumers or secondary producers—herbivores. Organisms which feed on microbes (bacteria and fungi) are termed microbivores. Animals that feed on primary consumers—carnivores—are secondary consumers. Each of these constitutes a trophic level. The sequence of consumption—from plant to herbivore, to carnivore—forms a food chain. Real systems are much more complex than this—organisms will generally feed on more than one form of food, and may feed at more than one trophic level. Carnivores may capture some prey which are part of a plant-based trophic system and others that are part of a detritus-based trophic system (a bird that feeds both on herbivorous grasshoppers and earthworms, which consume detritus). Real systems, with all these complexities, form food webs rather than food chains.

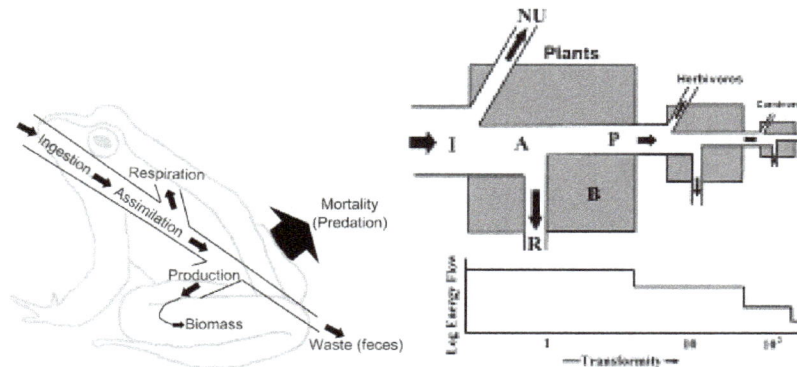

Left: Energy flow diagram of a frog. The frog represents a node in an extended food web. The energy ingested is utilized for metabolic processes and transformed into biomass. The energy flow continues on its path if the frog is ingested by predators, parasites, or as a decaying carcass in soil. This energy flow diagram illustrates how energy is lost as it fuels the metabolic process that transforms the energy and nutrients into biomass.
Right: An expanded three link energy food chain (1. plants, 2. herbivores, 3. carnivores) illustrating the relationship between food flow diagrams and energy transformity. The transformity of energy becomes degraded, dispersed, and diminished from higher quality to lesser quantity as the energy within a food chain flows from one trophic species into another. Abbreviations: I=input, A=assimilation, R=respiration, NU=not utilized, P=production, B=biomass.

Decomposition

The carbon and nutrients in dead organic matter are broken down by a group of processes known as decomposition. This releases nutrients that can then be re-used for plant and microbial production, and returns carbon dioxide to the atmosphere (or water) where it can be used for photosynthesis. In the absence of decomposition, dead organic matter would accumulate in an ecosystem and nutrients and atmospheric carbon dioxide would be depleted. Approximately 90% of terrestrial NPP goes directly from plant to decomposer.

Decomposition processes can be separated into three categories—leaching, fragmentation and chemical alteration of dead material. As water moves through dead organic matter, it dissolves and carries with it the water-soluble components. These are then taken up by organisms in the soil, react with mineral soil, or are transported beyond the confines of the ecosystem (and are considered "lost" to it). Newly shed leaves and newly dead animals have high concentrations of water-soluble components, and include sugars, amino acids and mineral nutrients. Leaching is more important in wet environments, and much less important in dry ones.

Fragmentation processes break organic material into smaller pieces, exposing new surfaces for colonization by microbes. Freshly shed leaf litter may be inaccessible due to an outer layer of cuticle or bark, and cell contents are protected by a cell wall. Newly dead animals may be covered by an exoskeleton. Fragmentation processes, which break through these protective layers, accelerate the rate of microbial decomposition. Animals fragment detritus as they hunt for food, as does passage through the gut. Freeze-thaw cycles and cycles of wetting and drying also fragment dead material.

The chemical alteration of dead organic matter is primarily achieved through bacterial and fungal action. Fungal hyphae produce enzymes which can break through the tough outer structures surrounding dead plant material. They also produce enzymes which break down lignin, which allows to them access to both cell contents and to the nitrogen in the lignin. Fungi can transfer carbon and nitrogen through their hyphal networks and thus, unlike bacteria, are not dependent solely on locally available resources.

Decomposition rates vary among ecosystems. The rate of decomposition is governed by three sets of factors—the physical environment (temperature, moisture and soil properties), the quantity and quality of the dead material available to decomposers, and the nature of the microbial community itself. Temperature controls the rate of microbial respiration; the higher the temperature, the faster microbial decomposition occurs. It also affects soil moisture, which slows microbial growth and reduces leaching. Freeze-thaw cycles also affect decomposition—freezing temperatures kill soil microorganisms, which allows leaching to play a more important role in moving nutrients around. This can be especially important as the soil thaws in the Spring, creating a pulse of nutrients which become available.

Decomposition rates are low under very wet or very dry conditions. Decomposition rates are highest in wet, moist conditions with adequate levels of oxygen. Wet soils tend to become deficient in oxygen (this is especially true in wetlands), which slows microbial growth. In dry soils, decomposition slows as well, but bacteria continue to grow (albeit at a slower rate) even after soils become too dry to support plant growth. When the rains return and soils become wet, the osmotic gradient between the bacterial cells and the soil water causes the cells to gain water quickly. Under these conditions, many bacterial cells burst, releasing a pulse of nutrients. Decomposition rates also tend to be slower in acidic soils. Soils which are rich in clay minerals tend to have lower decomposition rates, and thus, higher levels of organic matter. The smaller particles of clay result in a larger surface area that can hold water. The higher the water content of a soil, the lower the oxygen content and consequently, the lower the rate of decomposition. Clay minerals also bind particles of organic material to their surface, making them less accessibly to microbes. Soil disturbance like tilling increase decomposition by increasing the amount of oxygen in the soil and by exposing new organic matter to soil microbes.

The quality and quantity of the material available to decomposers is another major factor that influences the rate of decomposition. Substances like sugars and amino acids decompose readily and are considered "labile". Cellulose and hemicellulose, which are broken down more slowly, are "moderately labile". Compounds which are more resistant to decay, like lignin or cutin, are considered "recalcitrant". Litter with a higher proportion of labile compounds decomposes much more rapidly than does litter with a higher proportion of recalcitrant material. Consequently, dead animals decompose more rapidly than dead leaves, which themselves decompose more rapidly than fallen branches. As organic material in the soil ages, its quality decreases. The more labile compounds decompose quickly, leaving an increasing proportion of recalcitrant material. Microbial cell walls also contain recalcitrant materials like chitin, and these also accumulate as the microbes die, further reducing the quality of older soil organic matter.

Nutrient Cycling

Ecosystems continually exchange energy and carbon with the wider environment; mineral nutrients, on the other hand, are mostly cycled back and forth between plants, animals, microbes and the soil. Most nitrogen enters ecosystems through biological nitrogen fixation, is deposited through precipitation, dust, gases or is applied as fertilizer. Since most terrestrial ecosystems are nitrogen-limited, nitrogen cycling is an important control on ecosystem production.

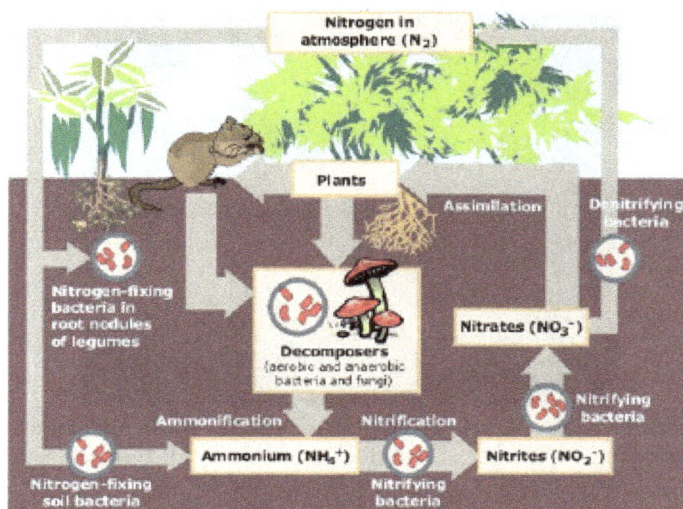

Biological nitrogen cycling

Until modern times, nitrogen fixation was the major source of nitrogen for ecosystems. Nitrogen fixing bacteria either live symbiotically with plants, or live freely in the soil. The energetic cost is high for plants which support nitrogen-fixing symbionts—as much as 25% of GPP when measured in controlled conditions. Many members of the legume plant family support nitrogen-fixing symbionts. Some cyanobacteria are also capable of nitrogen fixation. These are phototrophs, which carry out photosynthesis. Like other nitrogen-fixing bacteria, they can either be free-living or have symbiotic relationships with plants. Other sources of nitrogen include acid deposition produced through the combustion of fossil fuels, ammonia gas which evaporates from agricultural fields which have had fertilizers applied to them, and dust. Anthropogenic nitrogen inputs account for about 80% of all nitrogen fluxes in ecosystems.

When plant tissues are shed or are eaten, the nitrogen in those tissues becomes available to animals and microbes. Microbial decomposition releases nitrogen compounds from dead organic matter in the soil, where plants, fungi and bacteria compete for it. Some soil bacteria use organic nitrogen-containing compounds as a source of carbon, and release ammonium ions into the soil. This process is known as nitrogen mineralization. Others convert ammonium to nitrite and nitrate ions, a process known as nitrification. Nitric oxide and nitrous oxide are also produced during nitrification. Under nitrogen-rich and oxygen-poor conditions, nitrates and nitrites are converted to nitrogen gas, a process known as denitrification.

Other important nutrients include phosphorus, sulfur, calcium, potassium, magnesium and manganese. Phosphorus enters ecosystems through weathering. As ecosystems age this supply diminishes, making phosphorus-limitation more common in older landscapes (especially in the tropics). Calcium and sulfur are also produced by weathering, but acid deposition is an important source of sulfur in many ecosystems. Although magnesium and manganese are produced by weathering, exchanges between soil organic matter and living cells account for a significant portion of ecosystem fluxes. Potassium is primarily cycled between living cells and soil organic matter.

Function and Biodiversity

Loch Lomond in Scotland forms a relatively isolated ecosystem. The fish community of this lake has remained stable over a long period until a number of introductions in the 1970s restructured its food web.

Ecosystem processes are broad generalizations that actually take place through the actions of individual organisms. The nature of the organisms—the species, functional groups and trophic levels to which they belong—dictates the sorts of actions these individuals are capable of carrying out, and the relative efficiency with which they do so. Thus, ecosystem processes are driven by the number of species in an ecosystem, the exact nature of each individual species, and the relative abundance organisms within these species. Biodiversity plays an important role in ecosystem functioning.

Spiny forest at Ifaty, Madagascar, featuring various *Adansonia* (baobab) species, *Alluaudia procera* (Madagascar ocotillo) and other vegetation.

Ecological theory suggests that in order to coexist, species must have some level of limiting similarity—they must be different from one another in some fundamental way, otherwise one species would competitively exclude the other. Despite this, the cumulative effect of additional species in an ecosystem is not linear—additional species may enhance nitrogen retention, for example, but beyond some level of species richness, additional species may have little additive effect. The addition (or loss) of species which are ecologically similar to those already present in an ecosystem tends to only have a small effect on ecosystem function. Ecologically distinct species, on the other hand, have a much larger effect. Similarly, dominant species have a large impact on ecosystem function, while rare species tend to have a small effect. Keystone species tend to have an effect on ecosystem function that is disproportionate to their abundance in an ecosystem.

Ecosystem Goods and Services

Ecosystems provide a variety of goods and services upon which people depend. Ecosystem goods include the "tangible, material products" of ecosystem processes—food, construction material, medicinal plants—in addition to less tangible items like tourism and recreation, and genes from wild plants and animals that can be used to improve domestic species. Ecosystem services, on the other hand, are generally "improvements in the condition or location of things of value". These include things like the maintenance of hydrological cycles, cleaning air and water, the maintenance of oxygen in the atmosphere, crop pollination and even things like beauty, inspiration and opportunities for research. While ecosystem goods have traditionally been recognized as being the basis for things of economic value, ecosystem services tend to be taken for granted. While Gretchen Daily's original definition distinguished between ecosystem goods and ecosystem services, Robert Costanza and colleagues' later work and that of the Millennium Ecosystem Assessment lumped all of these together as ecosystem services.

Ecosystem Management

When natural resource management is applied to whole ecosystems, rather than single species, it is termed ecosystem management. A variety of definitions exist: F. Stuart Chapin and coauthors define it as "the application of ecological science to resource management to promote long-term sustainability of ecosystems and the delivery of essential ecosystem goods and services", while Norman Christensen and coauthors defined it as "management driven by explicit goals, executed by policies, protocols, and practices, and made adaptable by monitoring and research based on our best understanding of the ecological interactions and processes necessary to sustain ecosystem structure and function" and Peter Brussard and colleagues defined it as "managing areas at various scales in such a way that ecosystem services and biological resources are preserved while appropriate human use and options for livelihood are sustained".

Although definitions of ecosystem management abound, there is a common set of principles which underlie these definitions. A fundamental principle is the long-term sustainability of the production of goods and services by the ecosystem; "intergenerational sustainability [is] a precondition for management, not an afterthought". It also requires clear goals with respect to future trajectories and behaviors of the system being managed. Other important requirements include a sound ecological understanding of the system, including connectedness, ecological dynamics and the context in which the system is embedded. Other important principles include an understanding of the role of humans as components of the ecosystems and the use of adaptive management. While ecosystem management can be used as part of a plan for wilderness conservation, it can also be used in intensively managed ecosystems (for example, agroecosystem and close to nature forestry).

Ecosystem Dynamics

Ecosystems are dynamic entities—invariably, they are subject to periodic disturbances and are in the process of recovering from some past disturbance. When an ecosystem is subject to some sort of perturbation, it responds by moving away from its initial state. The tendency of a system to remain close to its equilibrium state, despite that disturbance, is termed its resistance. On the other hand, the speed with which it returns to its initial state after disturbance is called its resilience.

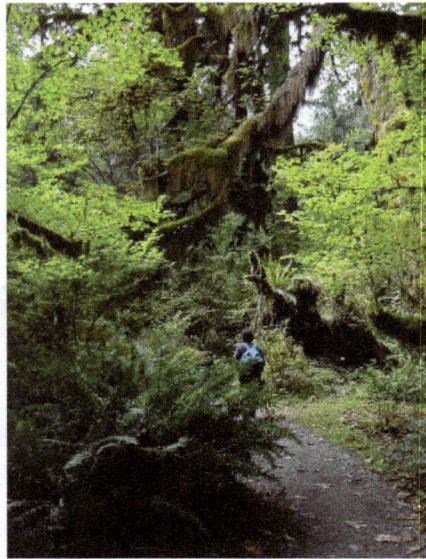

Temperate rainforest on the Olympic Peninsula in Washington state.

From one year to another, ecosystems experience variation in their biotic and abiotic environments. A drought, an especially cold winter and a pest outbreak all constitute short-term variability in environmental conditions. Animal populations vary from year to year, building up during resource-rich periods and crashing as they overshoot their food supply. These changes play out in changes in NPP, decomposition rates, and other ecosystem processes. Longer-term changes also shape ecosystem processes—the forests of eastern North America still show legacies of cultivation which ceased 200 years ago, while methane production in eastern Siberian lakes is controlled by organic matter which accumulated during the Pleistocene.

The High Peaks Wilderness Area in the 6,000,000-acre (2,400,000 ha) Adirondack Park is an example of a diverse ecosystem.

Disturbance also plays an important role in ecological processes. F. Stuart Chapin and coauthors define disturbance as "a relatively discrete event in time and space that alters the structure of populations, communities and ecosystems and causes changes in resources availability or the physical environment". This can range from tree falls and insect outbreaks to hurricanes and wildfires to volcanic eruptions and can cause large changes in plant, animal and microbe populations, as well soil organic matter content. Disturbance is followed by succession, a "directional change in ecosystem structure and functioning resulting from biotically driven changes in resources supply."

The frequency and severity of disturbance determines the way it impacts ecosystem function. Major disturbance like a volcanic eruption or glacial advance and retreat leave behind soils that lack plants, animals or organic matter. Ecosystems that experience disturbances that undergo primary succession. Less severe disturbance like forest fires, hurricanes or cultivation result in secondary succession. More severe disturbance and more frequent disturbance result in longer recovery times. Ecosystems recover more quickly from less severe disturbance events.

The early stages of primary succession are dominated by species with small propagules (seed and spores) which can be dispersed long distances. The early colonizers—often algae, cyanobacteria and lichens—stabilize the substrate. Nitrogen supplies are limited in new soils, and nitrogen-fixing species tend to play an important role early in primary succession. Unlike in primary succession, the species that dominate secondary succession, are usually present from the start of the process, often in the soil seed bank. In some systems the successional pathways are fairly consistent, and thus, are easy to predict. In others, there are many possible pathways—for example, the introduced nitrogen-fixing legume, *Myrica faya*, alter successional trajectories in Hawaiian forests.

The theoretical ecologist Robert Ulanowicz has used information theory tools to describe the structure of ecosystems, emphasizing mutual information (correlations) in studied systems. Drawing on this methodology and prior observations of complex ecosystems, Ulanowicz depicts approaches to determining the stress levels on ecosystems and predicting system reactions to defined types of alteration in their settings (such as increased or reduced energy flow, and eutrophication.

Ecosystem Ecology

Ecosystem ecology studies "the flow of energy and materials through organisms and the physical environment". It seeks to understand the processes which govern the stocks of material and energy in ecosystems, and the flow of matter and energy through them. The study of ecosystems can cover 10 orders of magnitude, from the surface layers of rocks to the surface of the planet.

There is no single definition of what constitutes an ecosystem. German ecologist Ernst-Detlef Schulze and coauthors defined an ecosystem as an area which is "uniform regarding the biological turnover, and contains all the fluxes above and below the ground area under consideration." They explicitly reject Gene Likens' use of entire river catchments as "too wide a demarcation" to be a single ecosystem, given the level of heterogeneity within such an area. Other authors have suggested that an ecosystem can encompass a much larger area, even the whole planet. Schulze and coauthors also rejected the idea that a single rotting log could be studied as an ecosystem because the size of the flows between the log and its surroundings are too large, relative to the proportion cycles within the log. Philosopher of science Mark Sagoff considers the failure to define "the kind of object it studies" to be an obstacle to the development of theory in ecosystem ecology.

A hydrothermal vent is an ecosystem on the ocean floor. (The scale bar is 1 m.)

Ecosystems can be studied through a variety of approaches—theoretical studies, studies monitoring specific ecosystems over long periods of time, those that look at differences between ecosystems to elucidate how they work and direct manipulative experimentation. Studies can be carried out at a variety of scales, from microcosms and mesocosms which serve as simplified representations of ecosystems, through whole-ecosystem studies. American ecologist Stephen R. Carpenter has argued that microcosm experiments can be "irrelevant and diversionary" if they are not carried out in conjunction with field studies carried out at the ecosystem scale, because microcosm experiments often fail to accurately predict ecosystem-level dynamics.

The Hubbard Brook Ecosystem Study, established in the White Mountains, New Hampshire in 1963, was the first successful attempt to study an entire watershed as an ecosystem. The study used stream chemistry as a means of monitoring ecosystem properties, and developed a detailed biogeochemical model of the ecosystem. Long-term research at the site led to the discovery of acid rain in North America in 1972, and was able to document the consequent depletion of soil cations (especially calcium) over the next several decades.

Classification

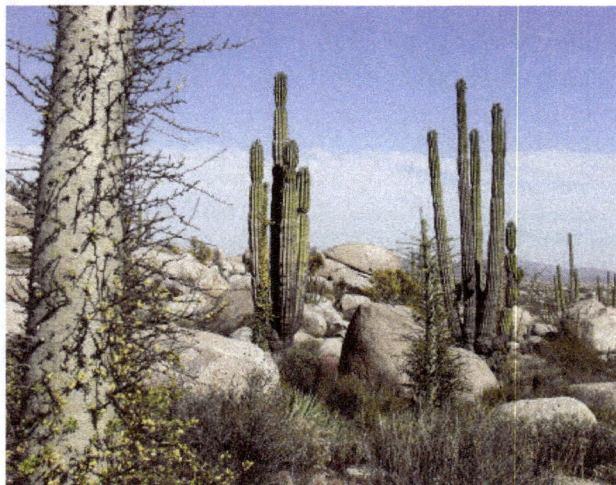

Flora of Baja California Desert, Cataviña region, Mexico.

Classifying ecosystems into ecologically homogeneous units is an important step towards effective ecosystem management. A variety of systems exist, based on vegetation cover, remote sensing, and bioclimatic classification systems. American geographer Robert Bailey defines a hierarchy of ecosystem units ranging from microecosystems (individual homogeneous sites, on the order of 10 square kilometres (4 sq mi) in area), through mesoecosystems (landscape mosaics, on the order of 1,000 square kilometres (400 sq mi)) to macroecosystems (ecoregions, on the order of 100,000 square kilometres (40,000 sq mi)).

Bailey outlined five different methods for identifying ecosystems: *gestalt* ("a whole that is not derived through considerable of its parts"), in which regions are recognized and boundaries drawn intuitively; a map overlay system where different layers like geology, landforms and soil types are overlain to identify ecosystems; multivariate clustering of site attributes; digital image processing of remotely sensed data grouping areas based on their appearance or other spectral properties; or by a "controlling factors method" where a subset of factors (like soils, climate, vegetation physiognomy or the distribution of plant or animal species) are selected from a large array of possible ones are used to delineate ecosystems. In contrast with Bailey's methodology, Puerto Rico ecologist Ariel Lugo and coauthors identified ten characteristics of an effective classification system: that it be based on georeferenced, quantitative data; that it should minimize subjectivity and explicitly identify criteria and assumptions; that it should be structured around the factors that drive ecosystem processes; that it should reflect the hierarchical nature of ecosystems; that it should be flexible enough to conform to the various scales at which ecosystem management operates; that it should be tied to reliable measures of climate so that it can "anticipat[e] global climate change; that it be applicable worldwide; that it should be validated against independent data; that it take into account the sometimes complex relationship between climate, vegetation and ecosystem functioning; and that it should be able to adapt and improve as new data become available".

Types

- Aquatic ecosystem
  - Marine ecosystem
    - Large marine ecosystem
  - Freshwater ecosystem
    - Lake ecosystem
    - River ecosystem
    - Wetland
- Terrestrial ecosystem
  - Forest
  - Littoral zone
  - Riparian zone
  - Subsurface lithoautotrophic microbial ecosystem

o   Urban ecosystem

o   Desert

A freshwater ecosystem in Gran Canaria, an island of the Canary Islands.

Anthropogenic Threats

As human populations and per capita consumption grow, so do the resource demands imposed on ecosystems and the impacts of the human ecological footprint. Natural resources are not invulnerable and infinitely available. The environmental impacts of anthropogenic actions, which are processes or materials derived from human activities, are becoming more apparent—air and water quality are increasingly compromised, oceans are being overfished, pests and diseases are extending beyond their historical boundaries, and deforestation is exacerbating flooding downstream. It has been reported that approximately 40–50% of Earth's ice-free land surface has been heavily transformed or degraded by anthropogenic activities, 66% of marine fisheries are either overexploited or at their limit, atmospheric CO_2 has increased more than 30% since the advent of industrialization, and nearly 25% of Earth's bird species have gone extinct in the last two thousand years. Society is increasingly becoming aware that ecosystem services are not only limited, but also that they are threatened by human activities. The need to better consider long-term ecosystem health and its role in enabling human habitation and economic activity is urgent. To help inform decision-makers, many ecosystem services are being assigned economic values, often based on the cost of replacement with anthropogenic alternatives. The ongoing challenge of prescribing economic value to nature, for example through biodiversity banking, is prompting transdisciplinary shifts in how we recognize and manage the environment, social responsibility, business opportunities, and our future as a species.

Alpha Diversity

In ecology, alpha diversity (α-diversity) is the mean species diversity in sites or habitats at a local scale. The term was introduced by R. H. Whittaker together with the terms beta diversity (β-diversity) and gamma diversity (γ-diversity). Whittaker's idea was that the total species diversity in a landscape (gamma diversity) is determined by two different things, the mean species diversity in sites or habitats at a more local scale (alpha diversity) and the differentiation among those habitats (beta diversity).

Scale Considerations

Both the area or landscape of interest and the sites or habitats within it may be of very different sizes in different situations, and no consensus has been reached on what spatial scales are appropriate to quantify alpha diversity. It has therefore been proposed that the definition of alpha diversity does not need to be tied to a specific spatial scale: alpha diversity can be measured for an existing dataset that consists of subunits at any scale. The subunits can be, for example, sampling units that were already used in the field when carrying out the inventory, or grid cells that are delimited just for the purpose of analysis. If results are extrapolated beyond the actual observations, it needs to be taken into account that the species diversity in the subunits generally gives an underestimation of the species diversity in larger areas.

Different Alpha Diversity Concepts

Ecologists have used several slightly different definitions of alpha diversity. Whittaker himself used the term both for the species diversity in a single subunit and for the mean species diversity in a collection of subunits. It has been argued that defining alpha diversity as a mean across all relevant subunits is preferable, because it agrees better with Whittaker's idea that total species diversity consists of alpha and beta components.

Definitions of alpha diversity can also differ in what they assume diversity to be. Often researchers use the values given by one or more diversity indices, such as species richness (richness is how many species and does not take into account the rarity of individuals, whereas biodiversity can), the Shannon index or the Simpson index. However, it has been argued that it would be better to use the effective number of species as the universal measure of species diversity. This measure allows weighting rare and abundant species in different ways, just as the diversity indices collectively do, but its meaning is intuitively easier to understand. The effective number of species is the number of equally-abundant species needed to obtain the same mean proportional species abundance as that observed in the dataset of interest (where all species may not be equally abundant).

Calculating Alpha Diversity

Suppose species diversity is equated with the effective number of species, and alpha diversity with the mean species diversity per subunit. Then alpha diversity can be calculated in two different ways that give the same result. The first approach is to calculate a weighted generalized mean of the within-subunit species proportional abundances, and then take the inverse of this mean. The second approach is to calculate the species diversity for each subunit separately, and then take a weighted generalized mean of these.

If the first approach is used, the equation is:

$$^{q}D_{\alpha} = \frac{1}{\sqrt[q-1]{\sum_{j=1}^{N}\sum_{i=1}^{S} p_{ij} p_{i|j}^{q-1}}}$$

In the equation, N is the total number of subunits and S is the total number of species (species richness) in the dataset. The proportional abundance of the ith species in the jth subunit is $p_{i|j}$. .These proportional abundances are weighted by the proportion of data that each contributes to the dataset, which equals p_{ij}. The denominator hence equals mean proportional species abundance within the subunits (mean $p_{i|j}$) as calculated with the weighted generalized mean with exponent q - 1.

If the second approach is used, the equation is:

$$^{q}D_{\alpha} = {}^{1-q}\!\sqrt{\sum_{j=1}^{N} w_{j}({}^{q}D_{\alpha j})^{1-q}}$$

This also equals a weighted generalized mean but with exponent 1 - q. Here the mean is taken of the $^{q}D_{\alpha j}$ values, each of which represents the effective species density (species diversity per subunit) in one subunit j. The nominal weight of the jth subunit is w_{j}, which equals the proportion of data that the subunit contributes to the dataset.

Large values of q lead to smaller alpha diversity than small values of q, because increasing q increases the effective weight given to those species with the highest proportional abundance and to those subunits with the lowest species diversity.

Examples Of Alpha Diversity

Alpha diversity can be calculated in both extinct and extant landscapes.

Examples of Extinct Alpha Diversity Studies

- The survival of amphibians and reptiles communities through the Permian-Triassic Extinction
- The reorganization of Ordovician benthic marine communities

Examples of Extant Alpha Diversity Studies

- High tree diversity in throughout the Amazon Rainforests of Ecuador

Extinction

The dodo of Mauritius, shown here in a 1626 illustration by Roelant Savery, is an often-cited example of modern extinction

In biology and ecology, extinction is the end of an organism or of a group of organisms (taxon), normally a species. The moment of extinction is generally considered to be the death of the last individual of the species, although the capacity to breed and recover may have been lost before this point. Because a species' potential range may be very large, determining this moment is difficult, and is usually done retrospectively. This difficulty leads to phenomena such as Lazarus taxa, where a species presumed extinct abruptly "reappears" (typically in the fossil record) after a period of apparent absence.

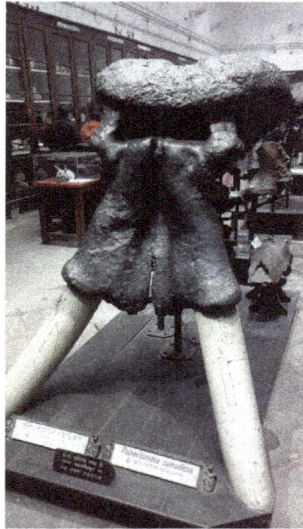

Skeleton of *Palaeoloxodon namadicus*, an extinct elephant species

Skeleton of various extinct dinosaurs; some other dinosaur lineages still flourish in the form of birds

More than 99 percent of all species, amounting to over five billion species, that ever lived on Earth are estimated to be extinct. Estimates on the number of Earth's current species range from 10 million to 14 million, of which about 1.2 million have been documented and over 86 percent have not yet been described.

Through evolution, species arise through the process of speciation—where new varieties of organisms arise and thrive when they are able to find and exploit an ecological niche—and species become extinct when they are no longer able to survive in changing conditions or against superior competition. The relationship between animals and their ecological niches has been firmly established. A typical species becomes extinct within 10 million years of its first appearance, although some species, called living fossils, survive with virtually no morphological change for hundreds of millions of years.

Mass extinctions are relatively rare events; however, isolated extinctions are quite common. Only recently have extinctions been recorded and scientists have become alarmed at the current high rate of extinctions. Most species that become extinct are never scientifically documented. Some scientists estimate that up to half of presently existing plant and animal species may become extinct by 2100.

Definition

A species is extinct when the last existing member dies. Extinction therefore becomes a certainty when there are no surviving individuals that can reproduce and create a new generation. A species may become functionally extinct when only a handful of individuals survive, which cannot reproduce due to poor health, age, sparse distribution over a large range, a lack of individuals of both sexes (in sexually reproducing species), or other reasons.

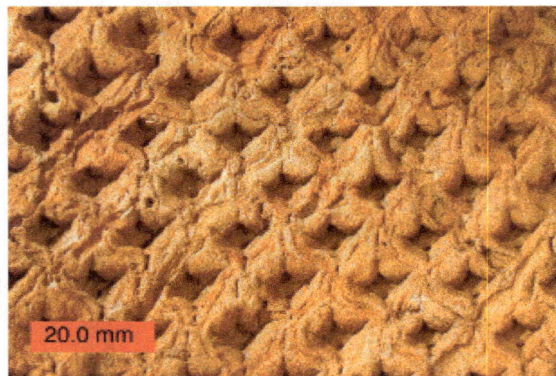

External mold of the extinct *Lepidodendron* from the Upper Carboniferous of Ohio

Pinpointing the extinction (or pseudoextinction) of a species requires a clear definition of that species. If it is to be declared extinct, the species in question must be uniquely distinguishable from any ancestor or daughter species, and from any other closely related species. Extinction of a species (or replacement by a daughter species) plays a key role in the punctuated equilibrium hypothesis of Stephen Jay Gould and Niles Eldredge.

In ecology, *extinction* is often used informally to refer to local extinction, in which a species ceases to exist in the chosen area of study, but may still exist elsewhere. This phenomenon is also known as extirpation. Local extinctions may be followed by a replacement of the species taken from other locations; wolf reintroduction is an example of this. Species which are not extinct are termed extant. Those that are extant but threatened by extinction are referred to as threatened or endangered species.

Currently an important aspect of extinction is human attempts to preserve critically endangered species. These are reflected by the creation of the conservation status "extinct in the wild" (EW). Species listed under this status by the International Union for Conservation of Nature (IUCN) are not known to have any living specimens in the wild, and are maintained only in zoos or other artificial environments. Some of these species are functionally extinct, as they are no longer part of their natural habitat and it is unlikely the species will ever be restored to the wild. When possible, modern zoological institutions try to maintain a viable population for species preservation and possible future reintroduction to the wild, through use of carefully planned breeding programs.

The extinction of one species' wild population can have knock-on effects, causing further extinctions. These are also called "chains of extinction". This is especially common with extinction of keystone species.

Pseudoextinction

Descendants may or may not exist for extinct species. Daughter species that evolve from a parent species carry on most of the parent species' genetic information, and even though the parent species may become extinct, the daughter species lives on. Extinction of a parent species where daughter species or subspecies are still extant is called pseudoextinction. In other cases, species have produced no new variants, or none that are able to survive the parent species' extinction.

Pseudoextinction is difficult to demonstrate unless one has a strong chain of evidence linking a living species to members of a pre-existing species. For example, it is sometimes claimed that the extinct *Hyracotherium*, which was an early horse that shares a common ancestor with the modern horse, is pseudoextinct, rather than extinct, because there are several extant species of *Equus*, including zebra and donkey. However, as fossil species typically leave no genetic material behind, one cannot say whether *Hyracotherium* evolved into more modern horse species or merely evolved from a common ancestor with modern horses. Pseudoextinction is much easier to demonstrate for larger taxonomic groups.

Lazarus Taxa

The coelacanth, a fish related to lungfish and tetrapods, was considered to have been extinct since the end of the Cretaceous Period until 1938 when a specimen was found, off the Chalumna River (now Tyolomnqa) on the east coast of South Africa. Museum curator Marjorie Courtenay-Latimer discovered the fish among the catch of a local angler, Captain Hendrick Goosen, on December 23, 1938. A local chemistry professor, JLB Smith, confirmed the fish's importance with a famous cable: "MOST IMPORTANT PRESERVE SKELETON AND GILLS = FISH DESCRIBED".

Far more recent possible or presumed extinctions of species which may turn out still to exist include the thylacine, or Tasmanian tiger (*Thylacinus cynocephalus*), the last known example of which died in Hobart Zoo in Tasmania in 1936; the Japanese wolf (*Canis lupus hodophilax*), last sighted over 100 years ago; the ivory-billed woodpecker (*Campephilus principalis*), last sighted for certain in 1944; and the slender-billed curlew (*Numenius tenuirostris*), not seen since 2007.

Causes

As long as species have been evolving, species have been going extinct. It is estimated that over 99.9% of all species that ever lived are extinct. The average life-span of a species is 1-10 million years, although this varies widely between taxa. There are a variety of causes that can contribute directly or indirectly to the extinction of a species or group of species. "Just as each species is unique", write Beverly and Stephen C. Stearns, "so is each extinction ... the causes for each are varied—some subtle and complex, others obvious and simple". Most simply, any species that cannot survive and reproduce in its environment and cannot move to a new environment where it can do so, dies out and becomes extinct. Extinction of a species may come suddenly when an otherwise healthy species is wiped out completely, as when toxic pollution renders its entire habitat unliveable; or may occur gradually over thousands or millions of years, such as when a species gradually loses out in competition for food to better adapted competitors. Extinction may occur a long time after the events that set it in motion, a phenomenon known as extinction debt.

The passenger pigeon, one of hundreds of species of extinct birds, was hunted to extinction over the course of a few decades

Assessing the relative importance of genetic factors compared to environmental ones as the causes of extinction has been compared to the debate on nature and nurture. The question of whether more extinctions in the fossil record have been caused by evolution or by catastrophe is a subject of discussion; Mark Newman, the author of *Modeling Extinction*, argues for a mathematical model that falls between the two positions. By contrast, conservation biology uses the extinction vortex model to classify extinctions by cause. When concerns about human extinction have been raised, for example in Sir Martin Rees' 2003 book *Our Final Hour*, those concerns lie with the effects of climate change or technological disaster.

Currently, environmental groups and some governments are concerned with the extinction of species caused by humanity, and they try to prevent further extinctions through a variety of conservation programs. Humans can cause extinction of a species through overharvesting, pollution, habitat destruction, introduction of invasive species (such as new predators and food competitors), overhunting, and other influences. Explosive, unsustainable human population growth is an essential cause of the extinction crisis. According to the International Union for Conservation of Nature (IUCN), 784 extinctions have been recorded since the year 1500, the arbitrary date selected to define "recent" extinctions, up to the year 2004; with many more likely to have gone unnoticed. Several species have also been listed as extinct since 2004.

Genetics and Demographic Phenomena

Population genetics and demographic phenomena affect the evolution, and therefore the risk of extinction, of species. Limited geographic range is the most important determinant of genus extinction at background rates but becomes increasingly irrelevant as mass extinction arises.

Natural selection acts to propagate beneficial genetic traits and eliminate weaknesses. But a dele-terious mutation can also be spread throughout a population through genetic drift.

Because traits are selected and not genes, the relationship between genetic diversity and extinction risk can be complex: factors such as balancing selection, cryptic genetic variation, phenotypic plas-ticity, and degeneracy all potentially play roles.

A diverse or deep gene pool gives a population a higher chance of surviving an adverse change in conditions. Effects that cause or reward a loss in genetic diversity can increase the chances of ex-tinction of a species. Population bottlenecks can dramatically reduce genetic diversity by severely limiting the number of reproducing individuals and make inbreeding more frequent. The founder effect can cause rapid, individual-based speciation and is the most dramatic example of a popula-tion bottleneck.

Genetic Pollution

Purebred wild species evolved to a specific ecology can be threatened with extinction through the process of genetic pollution—i.e., uncontrolled hybridization, introgression genetic swamping which leads to homogenization or out-competition from the introduced (or hybrid) species. Endemic populations can face such extinctions when new populations are imported or selectively bred by people, or when habitat modification brings previously isolated species into contact. Extinction is likeliest for rare species coming into contact with more abundant ones; interbreeding can swamp the rarer gene pool and create hybrids, depleting the purebred gene pool (for example, the endangered wild water buffalo is most threatened with extinction by genetic pollution from the abundant domestic water buffalo). Such extinctions are not al-ways apparent from morphological (non-genetic) observations. Some degree of gene flow is a normal evolutionarily process, nevertheless, hybridization (with or without introgression) threatens rare species' existence.

The gene pool of a species or a population is the variety of genetic information in its living members. A large gene pool (extensive genetic diversity) is associated with robust populations that can survive bouts of intense selection. Meanwhile, low genetic diversity reduces the range of adaptions possible. Replacing native with alien genes narrows genetic diversity within the original population, thereby increasing the chance of extinction.

Scorched land resulting from slash-and-burn agriculture

Habitat Degradation

Habitat degradation is currently the main anthropogenic cause of species extinctions. The main cause of habitat degradation worldwide is agriculture, with urban sprawl, logging, mining and some fishing practices close behind. The degradation of a species' habitat may alter the fitness landscape to such an extent that the species is no longer able to survive and becomes extinct. This may occur by direct effects, such as the environment becoming toxic, or indirectly, by limiting a species' ability to compete effectively for diminished resources or against new competitor species.

Habitat degradation through toxicity can kill off a species very rapidly, by killing all living members through contamination or sterilizing them. It can also occur over longer periods at lower toxicity levels by affecting life span, reproductive capacity, or competitiveness.

Habitat degradation can also take the form of a physical destruction of niche habitats. The widespread destruction of tropical rainforests and replacement with open pastureland is widely cited as an example of this; elimination of the dense forest eliminated the infrastructure needed by many species to survive. For example, a fern that depends on dense shade for protection from direct sunlight can no longer survive without forest to shelter it. Another example is the destruction of ocean floors by bottom trawling.

Diminished resources or introduction of new competitor species also often accompany habitat degradation. Global warming has allowed some species to expand their range, bringing unwelcome competition to other species that previously occupied that area. Sometimes these new competitors are predators and directly affect prey species, while at other times they may merely outcompete vulnerable species for limited resources. Vital resources including water and food can also be limited during habitat degradation, leading to extinction.

The golden toad was last seen on May 15, 1989. Decline in amphibian populations is ongoing worldwide

Predation, Competition, and Disease

In the natural course of events, species become extinct for a number of reasons, including but not limited to: extinction of a necessary host, prey or pollinator, inter-species competition, inability to deal with evolving diseases and changing environmental conditions (particularly sudden changes) which can act to introduce novel predators, or to remove prey. Recently in geological time, humans have become an additional cause of extinction (many people would say premature extinction) of some species, either as a new mega-predator or by transporting animals and plants from one part of the world to another. Such introductions have been occurring for thousands of years, sometimes intentionally (e.g. livestock released by sailors on islands as a future source of food) and sometimes accidentally (e.g. rats escaping from boats). In most cases, the introductions are unsuccessful, but when an invasive alien species does become established, the consequences can be catastrophic. Invasive alien species can affect native species directly by eating them, competing with them, and introducing pathogens or parasites that sicken or kill them; or indirectly by destroying or degrading their habitat. Human populations may themselves act as invasive predators. According to the "overkill hypothesis", the swift extinction of the megafauna in areas such as Australia (40,000 years before present), North and South America (12,000 years before present), Madagascar, Hawaii (300–1000 CE), and New Zealand (1300–1500 CE), resulted from the sudden introduction of human beings to environments full of animals that had never seen them before, and were therefore completely unadapted to their predation techniques.

Coextinction

Coextinction refers to the loss of a species due to the extinction of another; for example, the extinction of parasitic insects following the loss of their hosts. Coextinction can also occur when a species loses its pollinator, or to predators in a food chain who lose their prey. "Species coextinction is a manifestation of the interconnectedness of organisms in complex ecosystems ... While coextinction may not be the most important cause of species extinctions, it is certainly an insidious one". Coextinction is especially common when a keystone species goes extinct. Models suggest that coextinction is the most common form of biodiversity loss. There may be a cascade of coextinction across the trophic levels. Such effects are most severe in mutualistic and parasitic relationships. An example of coextinction is the Haast's eagle and the moa: the Haast's eagle was a predator that became extinct because its food source became extinct. The moa were several species of flightless birds that were a food source for the Haast's eagle.

The large Haast's eagle and moa from New Zealand

Climate Change

Extinction as a result of climate change has been confirmed by fossil studies. Particularly, the extinction of amphibians during the Carboniferous Rainforest Collapse, 305 million years ago. A 2003 review across 14 biodiversity research centers predicted that, because of climate change, 15–37% of land species would be "committed to extinction" by 2050. The ecologically rich areas that would potentially suffer the heaviest losses include the Cape Floristic Region, and the Caribbean Basin. These areas might see a doubling of present carbon dioxide levels and rising temperatures that could eliminate 56,000 plant and 3,700 animal species.

Mass Extinctions

The blue graph shows the apparent *percentage* (not the absolute number) of marine animal genera becoming extinct during any given time interval. It does not represent all marine species, just those that are readily fossilized.

There have been at least five mass extinctions in the history of life on earth, and four in the last 350 million years in which many species have disappeared in a relatively short period of geological time. A massive eruptive event is considered to be one likely cause of the "Permian–Triassic extinction event" about 250 million years ago, which is estimated to have killed 90% of species then existing. There is also evidence to suggest that this event was preceded by another mass extinction, known as Olson's Extinction. The Cretaceous–Paleogene extinction event (K-Pg) occurred 66 million years ago, at the end of the Cretaceous period, and is best known for having wiped out non-avian dinosaurs, among many other species.

Modern Extinctions

According to a 1998 survey of 400 biologists conducted by New York's American Museum of Natural History, nearly 70% believed that the Earth is currently in the early stages of a human-caused mass extinction, known as the Holocene extinction. In that survey, the same proportion of respondents agreed with the prediction that up to 20% of all living populations could become extinct within 30 years (by 2028). Biologist E. O. Wilson estimated in 2002 that if current rates of human destruction of the biosphere continue, one-half of all plant and animal species of life on earth will be extinct in 100 years. More significantly, the current rate of global species extinctions is estimated as 100 to 1000 times "background" rates (the average extinction rates in the evolutionary time scale of planet Earth), while future rates are likely 10,000 times higher. However, some groups are going extinct much faster.

History of Scientific Understanding

When it was first described in the 1750s, the idea of extinction was threatening to those who held a belief in the great chain of being, a theological position that did not allow for "missing links".

Dilophosaurus, one of the many extinct dinosaur genera. The cause of the Cretaceous–Paleogene extinction event is a subject of much debate amongst researchers

The possibility of extinction was not widely accepted before the 1800s. The devoted naturalist, Carl Linnaeus, could "hardly entertain" the idea that humans could cause the extinction of a species. When parts of the world had not been thoroughly examined and charted, scientists could not rule out that animals found only in the fossil record were not simply "hiding" in unexplored regions of the Earth. Georges Cuvier is credited with establishing extinction as a fact in a 1796 lecture to the French Institute. Cuvier's observations of fossil bones convinced him that they did not originate in extant animals. This discovery was critical for the spread of uniformitarianism, and led to the first book publicizing the idea of evolution though Cuvier himself strongly opposed the theories of evolution advanced by Lamarck and others.

Human Attitudes and Interests

Extinction is an important research topic in the field of zoology, and biology in general, and has also become an area of concern outside the scientific community. A number of organizations, such as the Worldwide Fund for Nature, have been created with the goal of preserving species from extinction. Governments have attempted, through enacting laws, to avoid habitat destruction, agricultural over-harvesting, and pollution. While many human-caused extinctions have been accidental, humans have also engaged in the deliberate destruction of some species, such as dangerous viruses, and the total destruction of other problematic species has been suggested. Other species were deliberately driven to extinction, or nearly so, due to poaching or because they were "undesirable", or to push for other human agendas. One example was the near extinction of the American bison, which was nearly wiped out by mass hunts sanctioned by the United States government, to force the removal of Native Americans, many of whom relied on the bison for food.

Biologist Bruce Walsh of the University of Arizona states three reasons for scientific interest in the preservation of species; genetic resources, ecosystem stability, and ethics; and today the scientific community "stress[es] the importance" of maintaining biodiversity.

In modern times, commercial and industrial interests often have to contend with the effects of production on plant and animal life. However, some technologies with minimal, or no, proven harmful effects on *Homo sapiens* can be devastating to wildlife (for example, DDT). Biogeographer Jared Diamond notes that while big business may label environmental concerns as "exaggerated", and often cause "devastating damage", some corporations find it in their interest to adopt good conservation practices, and even engage in preservation efforts that surpass those taken by national parks.

Governments sometimes see the loss of native species as a loss to ecotourism, and can enact laws with severe punishment against the trade in native species in an effort to prevent extinction in the wild. Nature preserves are created by governments as a means to provide continuing habitats to species crowded by human expansion. The 1992 Convention on Biological Diversity has resulted in international Biodiversity Action Plan programmes, which attempt to provide comprehensive guidelines for government biodiversity conservation. Advocacy groups, such as The Wildlands Project and the Alliance for Zero Extinctions, work to educate the public and pressure governments into action.

People who live close to nature can be dependent on the survival of all the species in their environment, leaving them highly exposed to extinction risks. However, people prioritize day-to-day survival over species conservation; with human overpopulation in tropical developing countries, there has been enormous pressure on forests due to subsistence agriculture, including slash-and-burn agricultural techniques that can reduce endangered species's habitats.

Planned Extinction

Completed

- The smallpox virus is now extinct in the wild, although samples are retained in laboratory settings.

- The rinderpest virus, which infected domestic cattle, is now extinct in the wild.

Proposed

The poliovirus is now confined to small parts of the world due to extermination efforts.

Dracunculus medinensis, a parasitic worm which causes the disease dracunculiasis, is now close to eradication thanks to efforts led by the Carter Center.

Treponema pallidum pertenue, a bacterium which causes the disease yaws, is in the process of being eradicated.

Biologist Olivia Judson has advocated the deliberate extinction of certain disease-carrying mosquito species. In a September 25, 2003 *New York Times* article, she advocated "specicide" of thirty mosquito species by introducing a genetic element which can insert itself into another crucial gene, to create recessive "knockout genes". She says that the *Anopheles* mosquitoes (which spread malaria) and *Aedes* mosquitoes (which spread dengue fever, yellow fever, elephantiasis, and other diseases) represent only 30 species; eradicating these would save at least one million human lives per annum, at a cost of reducing the genetic diversity of the family Culicidae by only 1%. She further argues that since species become extinct "all the time" the disappearance of a few more will not destroy the ecosystem: "We're not left with a wasteland every time a species vanishes. Removing one species sometimes causes shifts in the populations of other species—but different need not mean worse." In addition, anti-malarial and mosquito control programs offer little realistic hope to the 300 million people in developing nations who will be infected with acute illnesses this year. Although trials are ongoing, she writes that if they fail: "We should consider the ultimate swatting."

Cloning

Some, such as Harvard geneticist George M. Church, believe that ongoing technological advances will let us "bring back to life" an extinct species by cloning, using DNA from the remains of that species. Proposed targets for cloning include the mammoth, the thylacine, and the Pyrenean ibex. For this to succeed, enough individuals would have to be cloned, from the DNA of different individuals (in the case of sexually reproducing organisms) to create a viable population. Though bioethical and philosophical objections have been raised, the cloning of extinct creatures seems theoretically possible.

In 2003, scientists tried to clone the extinct Pyrenean ibex (*C. p. pyrenaica*). This attempt failed: of the 285 embryos reconstructed, 54 were transferred to 12 mountain goats and mountain goat-domestic goat hybrids, but only two survived the initial two months of gestation before they too died. In 2009, a second attempt was made to clone the Pyrenean ibex: one clone was born alive, but died seven minutes later, due to physical defects in the lungs.

Defaunation

Defaunation is the loss of animals from ecological communities. The growth of the human population, combined with advances in harvesting technologies, has led to more intense and efficient exploitation of the environment. This has resulted in the depletion of large vertebrates from ecological communities, creating what has been termed "empty forest." Defaunation differs from extinction; it includes both the disappearance of species and declines in abundance. Defaunation effects were first implied at the Symposium of Plant-Animal Interactions at the University of Campinas, Brazil in 1988 in the context of neotropical forests. Since then, the term has gained broader usage in conservation biology as a global phenomenon.

Drivers

Overexploitation

The intensive hunting and harvesting of animals threatens endangered vertebrate species across the world. Game vertebrates are considered valuable products of tropical forests and savannas. In Brazilian Amazonia, 23 million vertebrates are killed every year; large-bodied primates, tapirs, white-lipped peccaries, giant armadillos, and tortoises are some of the animals most sensitive to harvest. Overhunting can reduce the local population of such species by more than half, as well as reducing population density. Populations located nearer to villages are significantly more at risk of depletion. Abundance of local game species declines as density of local settlements, such as villages, increases.

Rhino poaching

Hunting and poaching may lead to local population declines or extinction in some species. Most affected species undergo pressure from multiple sources but the scientific community is still unsure of the complexity of these interactions and their feedback loops.

One case study in Panama found an inverse relationship between poaching intensity and abundance for 9 of 11 mammal species studied. In addition, preferred game species experienced greater declines and had higher spatial variation in abundance.

Habitat Destruction and Fragmentation

Human population growth results in changes in land-use, which can cause natural habitats to become fragmented, altered, or destroyed. Large mammals are often more vulnerable to extinction than smaller animals because they require larger home ranges and thus are more prone to suffer the effects of deforestation. Large species such as elephants, rhinoceroses, large primates, tapirs and peccaries are the first animals to disappear in fragmented rainforests.

Lacanja burn shows deforestation

A case study from Amazonian Ecuador analyzed two oil-road management approaches and their effects on the surrounding wildlife communities. The free-access road had forests that were cleared and fragmented and the other had enforced access control. Fewer species were found along the first road with density estimates being almost 80% lower than at the second site that which had minimal disturbance. This finding suggests that disturbances affected the local animals' willingness and ability to travel between patches.

Fragmentation lowers populations while increasing extinction risk when the remaining habitat size is small. When there is more unfragmented land, there is more habitat for more diverse species. A larger land patch also means it can accommodate more species with larger home ranges. However, when patch size decreases, there is an increase in the number of isolated fragments which can remain unoccupied by local fauna. If this persists, species may become extinct in the area.

A study on deforestation in the Amazon looked at two patterns of habitat fragmentation: "fish-bone" in smaller properties and another unnamed large property pattern. The large property pattern contained fewer fragments than the smaller fish-bone pattern. The results suggested that higher levels of fragmentation within the fish-bone pattern led to the loss of species and decreased diversity of large vertebrates. Human impacts, such as the fragmentation of forests, may cause large areas to lose the ability to maintain biodiversity and ecosystem function due to loss of key ecological processes. This can consequently cause changes within environments and skew evolutionary processes.

Invasive Species

Human influences, such as colonization and agriculture, have caused species to become distributed outside of their native ranges. Fragmentation also has cascading effects on native species, beyond reducing habitat and resource availability; it leaves areas vulnerable to non-native invasions. Invasive species can out-compete or directly prey upon native species, as well as alter the habitat so that native species can no longer survive.

In extinct animal species for which the cause of extinction is known, over 50% were affected by invasive species. For 20% of extinct animal species, invasive species are the only cited cause of extinction. Invasive species are the second-most important cause of extinction for mammals.

Global Patterns

Tropical regions are the most heavily impacted by defaunation. These regions, which include the Brazilian Amazon, the Congo Basin of Central Africa, and Indonesia, experience the greatest rates of overexploitation and habitat degradation. However, specific causes are varied, and areas with one endangered group (such as birds) do not necessarily also have other endangered groups (such as mammals, insects, or amphibians).

Deforestation of the Brazilian Amazon leads to habitat fragmentation and overexploitation. Hunting pressure in the Amazon rainforest has increased as traditional hunting techniques have been replaced by modern weapons such as shotguns. Access roads built for mining and logging operations fragment the forest landscape and allow hunters to move into forested areas which previously were untouched. The bushmeat trade in Central Africa incentivizes the overexploitation of local fauna. Indonesia has the most endangered animal species of any area in the world. International trade in wild animals, as well as extensive logging, mining and agriculture operations, drive the decline and extinction of numerous species.

Ecological Impacts

Genetic Loss

Inbreeding and genetic diversity loss often occur with endangered species populations because they have small and/or declining populations. Loss of genetic diversity lowers the ability of a population to deal with change in their environment and can make individuals within the community homogeneous. If this occurs, these animals are more susceptible to disease and other occurrences that may target a specific genome. Without genetic diversity, one disease could eradicate an entire species. Inbreeding lowers reproduction and survival rates. It is suggested that these genetic factors contribute to the extinction risk in threatened/endangered species.

Seed Dispersal

Effects on Plants and Forest Structure

The consequences of defaunation can be expected to affect the plant community. There are three non-mutually exclusive conclusions as to the consequences on tropical forest plant communities:

1. If seed dispersal agents are targeted by hunters, the effectiveness and amount of dispersal for those plant species will be reduced

2. The species composition of the seedling and sapling layers will be altered by hunting, and

3. Selective hunting of medium/large-sized animals instead of small-sized animals will lead to different seed predation patterns, with an emphasis on smaller seeds

One recent study analyzed seedling density and composition from two areas, Los Tuxtlas and Montes Azules. Los Tuxtlas, which is affected more by human activity, showed higher seedling density and a smaller average number of different species than in the other area. Results suggest that an absence of vertebrate dispersers can change the structure and diversity of forests. As a result, a plant community that relies on animals for dispersal could potentially have an altered biodiversity, species dominance, survival, demography, and spatial and genetic structure.

Poaching is likely to alter plant composition because the interactions between game and plant species varies in strength. Some game species interact strongly, weakly, or not at all with species. A change in plant species composition is likely to be a result because the net effect removal of game species varies among the plant species they interact with.

Effects on Small-bodied Seed Dispersers and Predators

As large-bodied vertebrates are increasingly lost from seed-dispersal networks, small-bodied seed dispersers (i.e. bats, birds, dung beetles) and seed predators (i.e. rodents) are affected. Defaunation leads to reduced species diversity. This is due to relaxed competition; small-bodied species normally compete with large-bodied vertebrates for food and other resources. As an area becomes defaunated, dominant small-bodied species take over, crowding out other similar species and leading to an overall reduced species diversity. The loss of species diversity is reflective of a larger loss of biodiversity, which has consequences for the maintenance of ecosystem services.

The quality of the physical habitat may also suffer. Bird and bat species (many of who are small bodied seed dispersers) rely on mineral licks as a source of sodium, which is not available elsewhere in their diets. In defaunated areas in the Western Amazon, mineral licks are more thickly covered by vegetation and have lower water availability. Bats were significantly less likely to visit these degraded mineral licks. The degradation of such licks will thus negatively affect the health and reproduction of bat populations.

Defaunation has negative consequences for seed dispersal networks as well. In the western Amazon, birds and bats have separate diets and thus form separate guilds within the network. It is hypothesized that large-bodied vertebrates, being generalists, connect separate guilds, creating a stable, resilient network. Defaunation results in a highly modular network in which specialized frugivores instead act as the connector hubs.

Ecosystem Services

Changes in predation dynamics, seed predation, seed dispersal, carrion removal, dung removal, vegetation trampling, and other ecosystem processes as a result of defaunation can affect ecosystem supporting and regulatory services, such as nutrient cycling and decomposition, crop pollination, pest control, and water quality.

Marine Defaunation

Defaunation in the ocean has occurred later and less intensely than on land. A relatively small number of marine species have been driven to extinction. However, many species have undergone local, ecological, and commercial extinction. Most large marine animal species still exist, such that the size distribution of global species assemblages has changed little since the Pleistocene, but individuals of each species are smaller on average, and overfishing has caused reductions in genetic diversity. Most extinctions and population declines to date have been driven by human overexploitation.

Consequences of Marine Defaunation

Marine defaunation has a wide array of effects on ecosystem structure and function. The loss of animals can have both top-down (cascading) and bottom-up effects, as well as consequences for biogeochemical cycling, ecosystem connectivity, and ecosystem stability.

Two of the most important ecosystem services threatened by marine defaunation are the provision of food and coastal storm protection.

Extinction Event

An extinction event (also known as a mass extinction or biotic crisis) is a widespread and rapid decrease in the biodiversity on Earth. Such an event is identified by a sharp change in the diversity and abundance of multicellular organisms. It occurs when the rate of extinction increases with respect to the rate of speciation. Because the majority of diversity and biomass on Earth is microbial, and thus difficult to measure, recorded extinction events affect the easily observed, biologically complex component of the biosphere rather than the total diversity and abundance of life.

Extinction occurs at an uneven rate. Based on the fossil record, the background rate of extinctions on Earth is about two to five taxonomic families of marine animals every million years. Marine fossils are mostly used to measure extinction rates because of their superior fossil record and stratigraphic range compared to land organisms.

The Great Oxygenation Event was probably the first major extinction event. Since the Cambrian explosion five further major mass extinctions have significantly exceeded the background extinction rate. The most recent and debatably best-known, the Cretaceous–Paleogene extinction event, which occurred approximately 66 million years ago (Ma), was a large-scale mass extinction of animal and plant species in a geologically short period of time. In addition to the five major mass extinctions, there are numerous minor ones as well, and the ongoing mass-extinction caused by human activity is sometimes called the sixth extinction. Mass extinctions seem to be a mainly Phanerozoic phenomenon, with extinction rates low before large complex organisms arose.

Estimates of the number of major mass extinctions in the last 540 million years range from as few as five to more than twenty. These differences stem from the threshold chosen for describing an extinction event as "major", and the data chosen to measure past diversity.

Major Extinction Events

In a landmark paper published in 1982, Jack Sepkoski and David M. Raup identified five mass extinctions. They were originally identified as outliers to a general trend of decreasing extinction rates during the Phanerozoic, but as more stringent statistical tests have been applied to the accumulating data, It has been established that multicellular animal life has experienced five major and many minor mass extinctions. The "Big Five" cannot be so clearly defined, but rather appear to represent the largest (or some of the largest) of a relatively smooth continuum of extinction events.

Badlands near Drumheller, Alberta, where erosion has exposed the K–Pg boundary

Trilobites were highly successful marine animals until the Permian–Triassic extinction event wiped them all out

1. Cretaceous–Paleogene extinction event (End Cretaceous, K-Pg extinction, or formerly K-T extinction): 66 Ma at the Cretaceous (Maastrichtian)-Paleogene (Danian) transition interval. The event formerly called the Cretaceous-Tertiary or K–T extinction or K-T boundary is now officially named the Cretaceous–Paleogene (or K–Pg) extinction event. About 17% of all families, 50% of all genera and 75% of all species became extinct. In the seas all the ammonites, plesiosaurs and mosasaurs disappeared and the percentage of sessile animals (those unable to move about) was reduced to about 33%. All non-avian dinosaurs became extinct during that time. The boundary event was severe with a significant amount of variability in the rate of extinction between and among different clades. Mammals and birds, the latter descended from theropod dinosaurs, emerged as dominant large land animals.

2. Triassic–Jurassic extinction event (End Triassic): 201.3 Ma at the Triassic-Jurassic transition. About 23% of all families, 48% of all genera (20% of marine families and 55% of marine genera) and 70% to 75% of all species became extinct. Most non-dinosaurian archosaurs, most therapsids, and most of the large amphibians were eliminated, leaving dinosaurs with little terrestrial competition. Non-dinosaurian archosaurs continued to dominate aquatic environments, while non-archosaurian diapsids continued to dominate marine environments. The Temnospondyl lineage of large amphibians also survived until the Cretaceous in Australia (e.g., *Koolasuchus*).

3. Permian–Triassic extinction event (End Permian): 252 Ma at the Permian-Triassic transition. Earth's largest extinction killed 57% of all families, 83% of all genera and 90% to 96% of all species (53% of marine families, 84% of marine genera, about 96% of all marine species and an estimated 70% of land species, including insects). The highly successful marine arthropod, the trilobite became extinct. The evidence of plants is less clear, but new taxa became dominant after the extinction. The "Great Dying" had enormous evolutionary significance: on land, it ended the primacy of mammal-like reptiles. The recovery of vertebrates took 30 million years, but the vacant niches created the opportunity for archosaurs to become ascendant. In the seas, the percentage of animals that were sessile dropped from 67% to 50%. The whole late Permian was a difficult time for at least marine life, even before the "Great Dying".

4. Late Devonian extinction: 375–360 Ma near the Devonian-Carboniferous transition. At the end of the Frasnian Age in the later part(s) of the Devonian Period, a prolonged series of extinctions eliminated about 19% of all families, 50% of all genera and at least 70% of all species. This extinction event lasted perhaps as long as 20 million years, and there is evidence for a series of extinction pulses within this period.

5. Ordovician–Silurian extinction events (End Ordovician or O-S): 450–440 Ma at the Ordovician-Silurian transition. Two events occurred that killed off 27% of all families, 57% of all genera and 60% to 70% of all species. Together they are ranked by many scientists as the second largest of the five major extinctions in Earth's history in terms of percentage of genera that became extinct.

Despite the popularization of these five events, there is no fine line separating them from other extinction events; using different methods of calculating an extinction's impact can lead to other events featuring in the top five.

The older the fossil record gets, the more difficult it is to read. This is because:

- Older fossils are harder to find as they are usually buried at a considerable depth.

- Dating older fossils is more difficult.

- Productive fossil beds are researched more than unproductive ones, therefore leaving certain periods unresearched.

- Prehistoric environmental events can disturb the deposition process.

- The preservation of fossils varies on land, but marine fossils tend to be better preserved than their sought after land-based counterparts.

It has been suggested that the apparent variations in marine biodiversity may actually be an artifact, with abundance estimates directly related to quantity of rock available for sampling from different time periods. However, statistical analysis shows that this can only account for 50% of the observed pattern, and other evidence (such as fungal spikes) provides reassurance that most widely accepted extinction events are real. A quantification of the rock exposure of Western Europe indicates that many of the minor events for which a biological explanation has been sought are most readily explained by sampling bias.

Research completed after the seminal 1982 paper has concluded that a sixth mass extinction event is ongoing:

> 6. Holocene extinction: Currently ongoing. Extinctions have occurred at over 1000 times the background extinction rate since 1900. The mass extinction is considered a result of human activity.

List of Extinction Events

This is a list of extinction events:

Period	Extinction	Date	Possible causes
Holocene	Holocene extinction event	c. 10,000 BCE — Ongoing	Humans
Pleistocene	Quaternary extinction event	640,000, 74,000, and 13,000 years ago	Unknown; may include climate changes and human overhunting
Pliocene	Pliocene–Pleistocene boundary marine extinction	2 Ma	Supernova? Eltanin impact?
Neogene	Middle Miocene disruption	14.5 Ma	
Palaeogene	Eocene–Oligocene extinction event	33.9 Ma	Popigai impactor?
	Cretaceous–Paleogene extinction event	66 Ma	Chicxulub impactor; Deccan Traps?
Cretaceous	Cenomanian-Turonian boundary event	94 Ma	Caribbean large igneous province
	Aptian extinction	117 Ma	
	End-Jurassic (Tithonian) extinction	145 Ma	
Jurassic	Toarcian turnover	183 Ma	Karoo-Ferrar Provinces
	Triassic–Jurassic extinction event	201 Ma	Central Atlantic magmatic province; impactor
Triassic	Carnian Pluvial Event	230 Ma	Wrangellia flood basalts
	Permian–Triassic extinction event	252 Ma	Siberian Traps; Wilkes Land Crater
Permian	Olson's Extinction	270 Ma	
Carboniferous	Carboniferous rainforest collapse	305 Ma	
Devonian	Late Devonian extinction	375-360 Ma	Viluy Traps
	End-Silurian extinction event	416 Ma	
	Lau event	420 Ma	Changes in sea level and chemistry?
	Mulde event	424 Ma	Global drop in sea level?
Silurian	Ireviken event	428 Ma	Deep-ocean anoxia; Milankovitch cycles?
	Ordovician–Silurian extinction events	450-440 Ma	Global cooling and sea level drop; Gamma-ray burst?
	Cambrian–Ordovician extinction events	488 Ma	

	Dresbachian extinction	502 Ma	
Cambrian	End-Botomian extinction event	517 Ma	
	End-Ediacaran extinction	542 Ma	
Precambrian	Great Oxygenation Event	2400 Ma	Rising oxygen levels in the atmosphere due to the development of photosynthesis

Evolutionary Importance

Mass extinctions have sometimes accelerated the evolution of life on Earth. When dominance of particular ecological niches passes from one group of organisms to another, it is rarely because the new dominant group is "superior" to the old and usually because an extinction event eliminates the old dominant group and makes way for the new one.

For example, mammaliformes ("almost mammals") and then mammals existed throughout the reign of the dinosaurs, but could not compete for the large terrestrial vertebrate niches which dinosaurs monopolized. The end-Cretaceous mass extinction removed the non-avian dinosaurs and made it possible for mammals to expand into the large terrestrial vertebrate niches. Ironically, the dinosaurs themselves had been beneficiaries of a previous mass extinction, the end-Triassic, which eliminated most of their chief rivals, the crurotarsans.

Another point of view put forward in the Escalation hypothesis predicts that species in ecological niches with more organism-to-organism conflict will be less likely to survive extinctions. This is because the very traits that keep a species numerous and viable under fairly static conditions become a burden once population levels fall among competing organisms during the dynamics of an extinction event.

Furthermore, many groups which survive mass extinctions do not recover in numbers or diversity, and many of these go into long-term decline, and these are often referred to as "Dead Clades Walking".

Darwin was firmly of the opinion that biotic interactions, such as competition for food and space—the 'struggle for existence'—were of considerably greater importance in promoting evolution and extinction than changes in the physical environment. He expressed this in *The Origin of Species*: "Species are produced and exterminated by slowly acting causes...and the most import of all causes of organic change is one which is almost independent of altered...physical conditions, namely the mutual relation of organism to organism-the improvement of one organism entailing the improvement or extermination of others".

Patterns in Frequency

It has been suggested variously that extinction events occurred periodically, every 26 to 30 million years, or that diversity fluctuates episodically every ~62 million years. Various ideas attempt to explain the supposed pattern, including the presence of a hypothetical companion star to the sun, oscillations in the galactic plane, or passage through the Milky Way's spiral arms. However, other authors have concluded the data on marine mass extinctions do not fit with the idea that mass extinctions are periodic, or that ecosystems gradually build up to a point at which a mass extinction is inevitable. Many of the proposed correlations have been argued to be spurious. Others have argued that there is strong evidence supporting periodicity in a variety of records, and additional evidence in the form of coincident periodic variation in nonbiological geochemical variables.

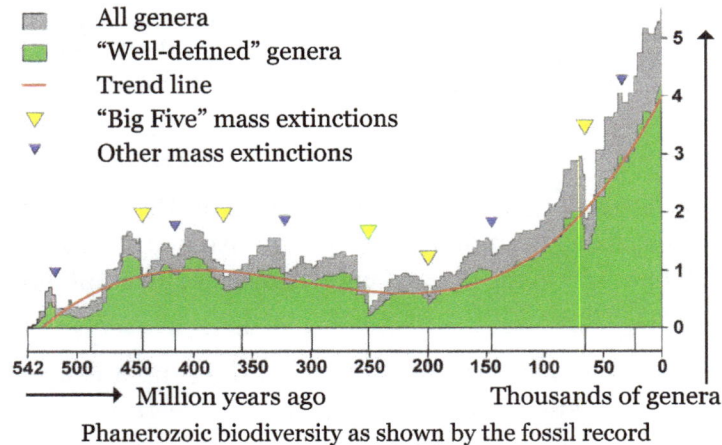

Phanerozoic biodiversity as shown by the fossil record

Mass extinctions are thought to result when a long-term stress is compounded by a short term shock. Over the course of the Phanerozoic, individual taxa appear to be less likely to become extinct at any time, which may reflect more robust food webs as well as less extinction-prone species and other factors such as continental distribution. However, even after accounting for sampling bias, there does appear to be a gradual decrease in extinction and origination rates during the Phanerozoic. This may represent the fact that groups with higher turnover rates are more likely to become extinct by chance; or it may be an artefact of taxonomy: families tend to become more speciose, therefore less prone to extinction, over time; and larger taxonomic groups (by definition) appear earlier in geological time.

It has also been suggested that the oceans have gradually become more hospitable to life over the last 500 million years, and thus less vulnerable to mass extinctions, but susceptibility to extinction at a taxonomic level does not appear to make mass extinctions more or less probable.

Causes

There is still debate about the causes of all mass extinctions. In general, large extinctions may result when a biosphere under long-term stress undergoes a short-term shock. An underlying mechanism appears to be present in the correlation of extinction and origination rates to diversity. High diversity leads to a persistent increase in extinction rate; low diversity to a persistent increase in origination rate. These presumably ecologically controlled relationships likely amplify smaller perturbations (asteroid impacts, etc.) to produce the global effects observed.

Identifying Causes of Particular Mass Extinctions

A good theory for a particular mass extinction should: (i) explain all of the losses, not just focus on a few groups (such as dinosaurs); (ii) explain why particular groups of organisms died out and why others survived; (iii) provide mechanisms which are strong enough to cause a mass extinction but not a total extinction; (iv) be based on events or processes that can be shown to have happened, not just inferred from the extinction.

It may be necessary to consider combinations of causes. For example, the marine aspect of the end-Cretaceous extinction appears to have been caused by several processes which partially overlapped in time and may have had different levels of significance in different parts of the world.

Arens and West (2006) proposed a "press / pulse" model in which mass extinctions generally require two types of cause: long-term pressure on the eco-system ("press") and a sudden catastrophe ("pulse") towards the end of the period of pressure. Their statistical analysis of marine extinction rates throughout the Phanerozoic suggested that neither long-term pressure alone nor a catastrophe alone was sufficient to cause a significant increase in the extinction rate.

Most Widely Supported Explanations

Macleod (2001) summarized the relationship between mass extinctions and events which are most often cited as causes of mass extinctions, using data from Courtillot *et al.* (1996), Hallam (1992) and Grieve *et al.* (1996):

- Flood basalt events: 11 occurrences, all associated with significant extinctions But Wignall (2001) concluded that only five of the major extinctions coincided with flood basalt eruptions and that the main phase of extinctions started before the eruptions.

- Sea-level falls: 12, of which seven were associated with significant extinctions.

- Asteroid impacts: one large impact is associated with a mass extinction, i.e. the Cretaceous–Paleogene extinction event; there have been many smaller impacts but they are not associated with significant extinctions.

The most commonly suggested causes of mass extinctions are listed below.

Flood Basalt Events

The formation of large igneous provinces by flood basalt events could have:

- produced dust and particulate aerosols which inhibited photosynthesis and thus caused food chains to collapse both on land and at sea

- emitted sulfur oxides which were precipitated as acid rain and poisoned many organisms, contributing further to the collapse of food chains

- emitted carbon dioxide and thus possibly causing sustained global warming once the dust and particulate aerosols dissipated.

Flood basalt events occur as pulses of activity punctuated by dormant periods. As a result, they are likely to cause the climate to oscillate between cooling and warming, but with an overall trend towards warming as the carbon dioxide they emit can stay in the atmosphere for hundreds of years.

It is speculated that massive volcanism caused or contributed to the End-Permian, End-Triassic and End-Cretaceous extinctions. The correlation between gigantic volcanic events expressed in the large igneous provinces and mass extinctions was shown for the last 260 Myr. Recently such possible correlation was extended for the whole Phanerozoic Eon.

Sea-level Falls

These are often clearly marked by worldwide sequences of contemporaneous sediments which show all or part of a transition from sea-bed to tidal zone to beach to dry land – and where there is no evidence that the rocks in the relevant areas were raised by geological processes such as orogeny. Sea-level falls could reduce the continental shelf area (the most productive part of the oceans) sufficiently to cause a marine mass extinction, and could disrupt weather patterns enough to cause extinctions on land. But sea-level falls are very probably the result of other events, such as sustained global cooling or the sinking of the mid-ocean ridges.

Sea-level falls are associated with most of the mass extinctions, including all of the "Big Five"— End-Ordovician, Late Devonian, End-Permian, End-Triassic, and End-Cretaceous.

A study, published in the journal Nature (online June 15, 2008) established a relationship between the speed of mass extinction events and changes in sea level and sediment. The study suggests changes in ocean environments related to sea level exert a driving influence on rates of extinction, and generally determine the composition of life in the oceans.

Impact Events

The impact of a sufficiently large asteroid or comet could have caused food chains to collapse both on land and at sea by producing dust and particulate aerosols and thus inhibiting photosynthesis. Impacts on sulfur-rich rocks could have emitted sulfur oxides precipitating as poisonous acid rain, contributing further to the collapse of food chains. Such impacts could also have caused megatsunamis and/or global forest fires.

Most paleontologists now agree that an asteroid did hit the Earth about 66 Ma ago, but there is an ongoing dispute whether the impact was the sole cause of the Cretaceous–Paleogene extinction event.

Sustained and Significant Global Cooling

Sustained global cooling could kill many polar and temperate species and force others to migrate towards the equator; reduce the area available for tropical species; often make the Earth's climate more arid on average, mainly by locking up more of the planet's water in ice and snow. The glaciation cycles of the current ice age are believed to have had only a very mild impact on biodiversity, so the mere existence of a significant cooling is not sufficient on its own to explain a mass extinction.

It has been suggested that global cooling caused or contributed to the End-Ordovician, Permian-Triassic, Late Devonian extinctions, and possibly others. Sustained global cooling is distinguished from the temporary climatic effects of flood basalt events or impacts.

Sustained and Significant Global Warming

This would have the opposite effects: expand the area available for tropical species; kill temperate species or force them to migrate towards the poles; possibly cause severe extinctions of polar species; often make the Earth's climate wetter on average, mainly by melting ice and snow and thus increasing the volume of the water cycle. It might also cause anoxic events in the oceans.

Global warming as a cause of mass extinction is supported by several recent studies.

The most dramatic example of sustained warming is the Paleocene-Eocene Thermal Maximum, which was associated with one of the smaller mass extinctions. It has also been suggested to have caused the Triassic-Jurassic extinction event, during which 20% of all marine families became extinct. Furthermore, the Permian–Triassic extinction event has been suggested to have been caused by warming.

Clathrate Gun Hypothesis

Clathrates are composites in which a lattice of one substance forms a cage around another. Methane clathrates (in which water molecules are the cage) form on continental shelves. These clathrates are likely to break up rapidly and release the methane if the temperature rises quickly or the pressure on them drops quickly—for example in response to sudden global warming or a sudden drop in sea level or even earthquakes. Methane is a much more powerful greenhouse gas than carbon dioxide, so a methane eruption ("clathrate gun") could cause rapid global warming or make it much more severe if the eruption was itself caused by global warming.

The most likely signature of such a methane eruption would be a sudden decrease in the ratio of carbon-13 to carbon-12 in sediments, since methane clathrates are low in carbon-13; but the change would have to be very large, as other events can also reduce the percentage of carbon-13.

It has been suggested that "clathrate gun" methane eruptions were involved in the end-Permian extinction ("the Great Dying") and in the Paleocene–Eocene Thermal Maximum, which was associated with one of the smaller mass extinctions.

Anoxic Events

Anoxic events are situations in which the middle and even the upper layers of the ocean become deficient or totally lacking in oxygen. Their causes are complex and controversial, but all known instances are associated with severe and sustained global warming, mostly caused by sustained massive volcanism.

It has been suggested that anoxic events caused or contributed to the Ordovician–Silurian, late Devonian, Permian–Triassic and Triassic–Jurassic extinctions, as well as a number of lesser extinctions (such as the Ireviken, Mulde, Lau, Toarcian and Cenomanian–Turonian events). On the other hand, there are widespread black shale beds from the mid-Cretaceous which indicate anoxic events but are not associated with mass extinctions.

The bio-availability of essential trace elements (in particular Selenium) to potentially lethal lows has been shown to coincide with, and likely have contributed to, at least three mass extinction events in the oceans, i.e. at the end of the Ordovician, during the Middle and Late Devonian, and at the end of the Triassic. During periods of low oxygen concentrations very soluble selenate (Se^{6+}) is converted into much less soluble selenide (Se^{2+}), elemental Se and organo-selenium complexes. Bio-availability of Selenium during these extinction events dropped to about 1% of the current oceanic concentration, a level that has been proven toxic to many extant organisms.

Hydrogen Sulfide Emissions from the Seas

Kump, Pavlov and Arthur (2005) have proposed that during the Permian–Triassic extinction event the warming also upset the oceanic balance between photosynthesising plankton and deep-water sulfate-reducing bacteria, causing massive emissions of hydrogen sulfide which poisoned life on both land and sea and severely weakened the ozone layer, exposing much of the life that still remained to fatal levels of UV radiation.

Oceanic Overturn

Oceanic overturn is a disruption of thermo-haline circulation which lets surface water (which is more saline than deep water because of evaporation) sink straight down, bringing anoxic deep water to the surface and therefore killing most of the oxygen-breathing organisms which inhabit the surface and middle depths. It may occur either at the beginning or the end of a glaciation, although an overturn at the start of a glaciation is more dangerous because the preceding warm period will have created a larger volume of anoxic water.

Unlike other oceanic catastrophes such as regressions (sea-level falls) and anoxic events, overturns do not leave easily identified "signatures" in rocks and are theoretical consequences of researchers' conclusions about other climatic and marine events.

It has been suggested that oceanic overturn caused or contributed to the late Devonian and Permian–Triassic extinctions.

A Nearby Nova, Supernova or Gamma Ray Burst

A nearby gamma-ray burst (less than 6000 light years away) would be powerful enough to destroy the Earth's ozone layer, leaving organisms vulnerable to ultraviolet radiation from the sun. Gamma ray bursts are fairly rare, occurring only a few times in a given galaxy per million years. It has been suggested that a supernova or gamma ray burst caused the End-Ordovician extinction.

Geomagnetic Reversal

One theory is that periods of increased geomagnetic reversals will weaken Earth's magnetic field long enough to expose the atmosphere to the solar winds, causing oxygen ions to escape the atmosphere in a rate increased by 3-4 orders, resulting in a disastrous drop of oxygen.

Plate Tectonics

Movement of the continents into some configurations can cause or contribute to extinctions in several ways: by initiating or ending ice ages; by changing ocean and wind currents and thus altering climate; by opening seaways or land bridges which expose previously isolated species to competition for which they are poorly adapted (for example, the extinction of most of South America's native ungulates and all of its large metatherians after the creation of a land bridge between North and South America). Occasionally continental drift creates a super-continent which includes the vast majority of Earth's land area, which in addition to the effects listed above is likely to reduce the total area of continental shelf (the most species-rich part of the ocean) and produce a vast, arid continental interior which may have extreme seasonal variations.

Another theory is that the creation of the super-continent Pangaea contributed to the End-Permian mass extinction. Pangaea was almost fully formed at the transition from mid-Permian to late-Permian, and the "Marine genus diversity" diagram at the top of this article shows a level of extinction starting at that time which might have qualified for inclusion in the "Big Five" if it were not overshadowed by the "Great Dying" at the end of the Permian.

Other Hypotheses

Many other hypotheses have been proposed, such as the spread of a new disease, or simple out-competition following an especially successful biological innovation. But all have been rejected, usually for one of the following reasons: they require events or processes for which there is no evidence; they assume mechanisms which are contrary to the available evidence; they are based on other theories which have been rejected or superseded.

Scientists have been concerned that human activities could cause more plants and animals to become extinct than any point in the past. Along with human-made changes in climate, some of these extinctions could be caused by overhunting, overfishing, invasive species, or habitat loss.

Future Biosphere Extinction

The eventual warming and expanding of the Sun, combined with the eventual decline of atmospheric carbon dioxide could actually cause an even greater mass extinction, having the potential to wipe out even microbes, where rising global temperatures caused by the expanding Sun will gradually increase the rate of weathering, which in turn removes more and more carbon dioxide from the atmosphere. When carbon dioxide levels get too low (perhaps at 50 ppm), all plant life will die out, although simpler plants like grasses and mosses can survive much longer, until CO_2 levels drop to 10 ppm.

With all photosynthetic organisms gone, atmospheric oxygen can no longer be replenished, and is eventually removed by chemical reactions in the atmosphere, perhaps from volcanic eruptions. Eventually the loss of oxygen will cause all remaining aerobic life to die out via asphyxiation, leaving behind only simple anaerobic prokaryotes. When the Sun becomes 10% brighter in about a billion years, Earth will suffer a moist greenhouse effect resulting in its oceans boiling away, while the Earth's liquid outer core freezes due to the inner core's expansion and causes the Earth's magnetic field to shut down. In the absence of a magnetic field, charged particles from the Sun will deplete the atmosphere and further increase the Earth's temperature to an average of ~420 K (147 °C, 296 °F) in 2.8 billion years, causing the last remaining life on Earth to die out. This is the most extreme instance of a climate-caused extinction event. Since this will only happen late in the Sun's life, such will cause the final mass extinction in Earth's history (albeit a very long extinction event).

Effects and Recovery

The impact of mass extinction events varied widely. After a major extinction event, usually only weedy species survive due to their ability to live in diverse habitats. Later, species diversify and occupy empty niches. Generally, biodiversity recovers 5 to 10 million years after the extinction event. In the most severe mass extinctions it may take 15 to 30 million years.

The worst event, the Permian–Triassic extinction event, devastated life on earth and is estimated to have killed off over 90% of species. Life seemed to recover quickly after the P-T extinction, but this was mostly in the form of disaster taxa, such as the hardy *Lystrosaurus*. The most recent research indicates that the specialized animals that formed complex ecosystems, with high biodiversity, complex food webs and a variety of niches, took much longer to recover. It is thought that this long recovery was due to the successive waves of extinction which inhibited recovery, as well as to prolonged environmental stress to organisms which continued into the Early Triassic. Recent research indicates that recovery did not begin until the start of the mid-Triassic, 4M to 6M years after the extinction; and some writers estimate that the recovery was not complete until 30M years after the P-Tr extinction, i.e. in the late Triassic. Subsequent to the PT mass extinction, there was an increase in provincialization, with species occupying smaller ranges - perhaps removing incumbents from niches and setting the stage for an eventual rediversification.

The effects of mass extinctions on plants are somewhat harder to quantify, given the biases inherent in the plant fossil record. Some mass extinctions (such as the end-Permian) were equally catastrophic for plants, whereas others, such as the end-Devonian, did not affect the flora.

Snowball Earth

The Snowball Earth hypothesis proposes that Earth's surface became entirely or nearly entirely frozen at least once, sometime earlier than 650 Mya (million years ago). Proponents of the hypothesis argue that it best explains sedimentary deposits generally regarded as of glacial origin at tropical paleolatitudes, and other otherwise enigmatic features in the geological record. Opponents of the hypothesis contest the implications of the geological evidence for global glaciation, the geophysical feasibility of an ice- or slush-covered ocean, and the difficulty of escaping an all-frozen condition. A number of unanswered questions exist, including whether Earth was a full snowball, or a "slushball" with a thin equatorial band of open (or seasonally open) water.

The geological time frames under consideration come before the sudden radiation of multicellular bioforms on Earth known as the Cambrian explosion, and the most recent snowball episode indeed may have triggered the evolution of multicellularity. Another, much earlier and longer snowball episode, the Huronian glaciation, which occurred 2400 to 2100 Mya, may have been triggered by the first appearance of oxygen in the atmosphere, the "Great Oxygenation Event."

History

Sir Douglas Mawson (1882–1958), an Australian geologist and Antarctic explorer, spent much of his career studying the Neoproterozoic stratigraphy of South Australia, where he identified thick and extensive glacial sediments and late in his career speculated about the possibility of global glaciation.

Mawson's ideas of global glaciation, however, were based on the mistaken assumption that the geographic position of Australia, and that of other continents where low-latitude glacial deposits are found, has remained constant through time. With the advancement of the continental drift hypothesis, and eventually plate tectonic theory, came an easier explanation for the glaciogenic sediments — they were deposited at a point in time when the continents were at higher latitudes.

In 1964, the idea of global-scale glaciation reemerged when W. Brian Harland published a paper in which he presented palaeomagnetic data showing that glacial tillites in Svalbard and Greenland were deposited at tropical latitudes. From this palaeomagnetic data, and the sedimentological evidence that the glacial sediments interrupt successions of rocks commonly associated with tropical to temperate latitudes, he argued for an ice age that was so extreme that it resulted in the deposition of marine glacial rocks in the tropics.

In the 1960s, Mikhail Budyko, a Russian climatologist, developed a simple energy-balance climate model to investigate the effect of ice cover on global climate. Using this model, Budyko found that if ice sheets advanced far enough out of the polar regions, a feedback loop ensued where the increased reflectiveness (albedo) of the ice led to further cooling and the formation of more ice, until the entire Earth was covered in ice and stabilized in a new ice-covered equilibrium. While Budyko's model showed that this ice-albedo stability could happen, he concluded that it had in fact never happened, because his model offered no way to escape from such a feedback loop. In 1971, Aron Faegre, an American physicist, showed that a similar energy-balance model predicted three stable global climates, one of which was snowball earth. This model introduced Edward Norton Lorenz's concept of intransitivity indicating that there could be a major jump from one climate to another, including to snowball earth.

The term "snowball Earth" was coined by Joseph Kirschvink in a short paper published in 1992 within a lengthy volume concerning the biology of the Proterozoic eon. The major contributions from this work were: (1) the recognition that the presence of banded iron formations is consistent with such a global glacial episode, and (2) the introduction of a mechanism by which to escape from a completely ice-covered Earth — specifically, the accumulation of CO_2 from volcanic outgassing leading to an ultra-greenhouse effect.

Franklyn Van Houten's discovery of a consistent geological pattern in which lake levels rose and fell is now known as the "Van Houten cycle." His studies of phosphorus deposits and banded iron formations in sedimentary rocks made him an early adherent of the "snowball Earth" hypothesis postulating that the planet's surface froze more than 650 million years ago.

Interest in the notion of a snowball Earth increased dramatically after Paul F. Hoffman and his co-workers applied Kirschvink's ideas to a succession of Neoproterozoic sedimentary rocks in Namibia and elaborated upon the hypothesis in the journal *Science* in 1998 by incorporating such observations as the occurrence of cap carbonates.

In 2010, Francis MacDonald reported evidence that Pangaea was at equatorial latitude during the Cryogenian period with glacial ice at or below sea level, and that the associated Sturtian glaciation was global.

Evidence

The snowball Earth hypothesis was originally devised to explain geological evidence for the apparent presence of glaciers at tropical latitudes. According to modelling, an ice-albedo feedback would result in glacial ice rapidly advancing to the equator once the glaciers spread to within 25° to 30° of the equator. Therefore, the presence of glacial deposits within the tropics suggests global ice cover.

Critical to an assessment of the validity of the theory, therefore, is an understanding of the reliability and significance of the evidence that led to the belief that ice ever reached the tropics. This evidence must prove two things:

1. that a bed contains sedimentary structures that could have been created only by glacial activity;

2. that the bed lay within the tropics when it was deposited.

During a period of global glaciation, it must also be demonstrated that glaciers were active at different global locations at the same time, and that no other deposits of the same age are in existence.

This last point is very difficult to prove. Before the Ediacaran, the biostratigraphic markers usually used to correlate rocks are absent; therefore there is no way to prove that rocks in different places across the globe were deposited at precisely the same time. The best that can be done is to estimate the age of the rocks using radiometric methods, which are rarely accurate to better than a million years or so.

The first two points are often the source of contention on a case-to-case basis. Many glacial features can also be created by non-glacial means, and estimating the approximate latitudes of landmasses even as recently as 200 million years ago can be riddled with difficulties.

Palaeomagnetism

The snowball Earth hypothesis was first posited in order to explain what were then considered to be glacial deposits near the equator. Since tectonic plates move slowly over time, ascertaining their position at a given point in Earth's long history is not easy. In addition to considerations of how the recognizable landmasses could have fit together, the latitude at which a rock was deposited can be constrained by palaeomagnetism.

When sedimentary rocks form, magnetic minerals within them tend to align themselves with the Earth's magnetic field. Through the precise measurement of this palaeomagnetism, it is possible to estimate the latitude (but not the longitude) where the rock matrix was deposited. Palaeomagnetic measurements have indicated that some sediments of glacial origin in the Neoproterozoic rock record were deposited within 10 degrees of the equator, although the accuracy of this reconstruction is in question. This palaeomagnetic location of apparently glacial sediments (such as dropstones) has been taken to suggest that glaciers extended to sea level in tropical latitudes at the time the sediments were deposited. It is not clear whether this implies a global glaciation, or the existence of localized, possibly land-locked, glacial regimes. Others have even suggested that most data do not constrain any glacial deposits to within 25° of the equator.

Skeptics suggest that the palaeomagnetic data could be corrupted if Earth's ancient magnetic field was substantially different from today's. Depending on the rate of cooling of Earth's core, it is possible that during the Proterozoic, the magnetic field did not approximate a simple dipolar distribution, with north and south magnetic poles roughly aligning with the planet's axis as they do today. Instead, a hotter core may have circulated more vigorously and given rise to 4, 8 or more poles. Palaeomagnetic data would then have to be re-interpreted, as the sedimentary minerals could have aligned pointing to a 'West Pole' rather than the North Pole. Alternatively, Earth's dipolar field could have been oriented such that the poles were close to the equator. This hypothesis has been posited to explain the extraordinarily rapid motion of the magnetic poles implied by the Ediacaran palaeomagnetic record; the alleged motion of the north pole would occur around the same time as the Gaskiers glaciation.

Another weakness of reliance on palaeomagnetic data is the difficulty in determining whether the magnetic signal recorded is original, or whether it has been reset by later activity. For example, a mountain-building orogeny releases hot water as a by-product of metamorphic reactions; this water can circulate to rocks thousands of kilometers away and reset their magnetic signature. This makes the authenticity of rocks older than a few million years difficult to determine without painstaking mineralogical observations. Moreover, further evidence is accumulating that large-scale remagnetization events have taken place which may necessitate revision of the estimated positions of the palaeomagnetic poles.

There is currently only one deposit, the Elatina deposit of Australia, that was indubitably deposited at low latitudes; its depositional date is well-constrained, and the signal is demonstrably original.

Low-latitude Glacial Deposits

Diamictite of the Neoproterozoic Pocatello Formation, a 'snowball Earth'–type deposit

Elatina Fm diamictite below Ediacaran GSSP site in the Flinders Ranges NP, South Australia. A$1 coin for scale.

Sedimentary rocks that are deposited by glaciers have distinctive features that enable their identification. Long before the advent of the *snowball Earth* hypothesis many Neoproterozoic sediments had been interpreted as having a glacial origin, including some apparently at tropical latitudes at the time of their deposition. However, it is worth remembering that many sedimentary features traditionally associated with glaciers can also be formed by other means. Thus the glacial origin of many of the key occurrences for snowball Earth has been contested. As of 2007, there was only one "very reliable" – still challenged – datum point identifying tropical tillites, which makes statements of equatorial ice cover somewhat presumptuous. However evidence of sea-level glaciation in the tropics during the Sturtian is accumulating. Evidence of possible glacial origin of sediment includes:

- Dropstones (stones dropped into marine sediments), which can be deposited by glaciers or other phenomena.

- Varves (annual sediment layers in periglacial lakes), which can form at higher temperatures.

- Glacial striations (formed by embedded rocks scraped against bedrock): similar striations are from time to time formed by mudflows or tectonic movements.

- Diamictites (poorly sorted conglomerates). Originally described as glacial till, most were in fact formed by debris flows.

Open-water Deposits

It appears that some deposits formed during the snowball period could only have formed in the presence of an active hydrological cycle. Bands of glacial deposits up to 5,500 meters thick, separated by small (meters) bands of non-glacial sediments, demonstrate that glaciers melted and re-formed repeatedly for tens of millions of years; solid oceans would not permit this scale of deposition. It is considered possible that ice streams such as seen in Antarctica today could have caused these sequences. Further, sedimentary features that could only form in open water (for example: wave-formed ripples, far-traveled ice-rafted debris and indicators of photosynthetic activity) can be found throughout sediments dating from the snowball-Earth periods. While these may represent "oases" of meltwater on a completely frozen Earth, computer modelling suggests that large areas of the ocean must have remained ice-free; arguing that a "hard" snowball is not plausible in terms of energy balance and general circulation models.

Carbon Isotope Ratios

There are two stable isotopes of carbon in sea water: carbon-12 (^{12}C) and the rare carbon-13 (^{13}C), which makes up about 1.109 percent of carbon atoms.

Biochemical processes, of which photosynthesis is one, tend to preferentially incorporate the lighter ^{12}C isotope. Thus ocean-dwelling photosynthesizers, both protists and algae, tend to be very slightly depleted in ^{13}C, relative to the abundance found in the primary volcanic sources of Earth's carbon. Therefore, an ocean with photosynthetic life will have a lower $^{13}C/^{12}C$ ratio within organic remains, and a lower ratio in corresponding ocean water. The organic component of the lithified sediments will forever remain very slightly, but measurably, depleted in ^{13}C.

During the proposed episode of snowball Earth, there are rapid and extreme negative excursions in the ratio of ^{13}C to ^{12}C. This is consistent with a deep freeze that killed off most or nearly all photosynthetic life – although other mechanisms, such as clathrate release, can also cause such perturbations. Close analysis of the timing of ^{13}C 'spikes' in deposits across the globe allows the recognition of four, possibly five, glacial events in the late Neoproterozoic.

Banded Iron Formations

Banded iron formations (BIF) are sedimentary rocks of layered iron oxide and iron-poor chert. In the presence of oxygen, iron naturally rusts and becomes insoluble in water. The banded iron formations are commonly very old and their deposition is often related to the oxidation of Earth's atmosphere during the Paleoproterozoic era, when dissolved iron in the ocean came in contact with photosynthetically produced oxygen and precipitated out as iron oxide.

2.1 billion year old rock with black-band ironstone

The bands were produced at the tipping point between an anoxic and an oxygenated ocean. Since today's atmosphere is oxygen rich (nearly 21% by volume) and in contact with the oceans, it is not possible to accumulate enough iron oxide to deposit a banded formation. The only extensive iron formations that were deposited after the Paleoproterozoic (after 1.8 billion years ago) are associated with Cryogenian glacial deposits.

For such iron-rich rocks to be deposited there would have to be anoxia in the ocean, so that much dissolved iron (as ferrous oxide) could accumulate before it met an oxidant that would precipitate it as ferric oxide. For the ocean to become anoxic it must have limited gas exchange with the oxygenated atmosphere. Proponents of the hypothesis argue that the reappearance of BIF in the sedimentary record is a result of limited oxygen levels in an ocean sealed by sea ice, while opponents suggest that the rarity of the BIF deposits may indicate that they formed in inland seas.

Being isolated from the oceans, such lakes may have been stagnant and anoxic at depth, much like today's Black Sea; a sufficient input of iron could provide the necessary conditions for BIF formation. A further difficulty in suggesting that BIFs marked the end of the glaciation is that they are found interbedded with glacial sediments. BIFs are also strikingly absent during the Marinoan glaciation.

Cap carbonate rocks

Around the top of Neoproterozoic glacial deposits there is commonly a sharp transition into a chemically precipitated sedimentary limestone or dolostone metres to tens of metres thick. These cap carbonates sometimes occur in sedimentary successions that have no other carbonate rocks, suggesting that their deposition is result of a profound aberration in ocean chemistry.

A present-day glacier

These cap carbonates have unusual chemical composition, as well as strange sedimentary structures that are often interpreted as large ripples. The formation of such sedimentary rocks could be caused by a large influx of positively charged ions, as would be produced by rapid weathering during the extreme greenhouse following a snowball Earth event. The $\delta^{13}C$ isotopic signature of the cap carbonates is near -5 ‰, consistent with the value of the mantle — such a low value is usually/could be taken to signify an absence of life, since photosynthesis usually acts to raise the value; alternatively the release of methane deposits could have lowered it from a higher value, and counterbalance the effects of photosynthesis.

Volcanoes may have had a role in replenishing CO_2, possibly ending the global ice age that was the snowball Earth during the Cryogenian Period.

The precise mechanism involved in the formation of cap carbonates is not clear, but the most cited explanation suggests that at the melting of a snowball Earth, water would dissolve the abundant CO_2 from the atmosphere to form carbonic acid, which would fall as acid rain. This would weather exposed silicate and carbonate rock (including readily attacked glacial debris), releasing large amounts of calcium, which when washed into the ocean would form distinctively textured layers of carbonate sedimentary rock. Such an abiotic "cap carbonate" sediment can be found on top of the glacial till that gave rise to the snowball Earth hypothesis.

However, there are some problems with the designation of a glacial origin to cap carbonates. Firstly, the high carbon dioxide concentration in the atmosphere would cause the oceans to become acidic, and dissolve any carbonates contained within — starkly at odds with the deposition of cap carbonates. Further, the thickness of some cap carbonates is far above what could reasonably be produced in the relatively quick deglaciations. The cause is further weakened by the lack of cap carbonates above many sequences of clear glacial origin at a similar time and the occurrence of similar carbonates within the sequences of proposed glacial origin. An alternative mechanism, which may have produced the Doushantuo cap carbonate at least, is the rapid, widespread release of methane. This accounts for incredibly low — as low as -48 ‰ — $\delta^{13}C$ values — as well as unusual sedimentary features which appear to have been formed by the flow of gas through the sediments.

Changing Acidity

Isotopes of the element boron suggest that the pH of the oceans dropped dramatically before and after the Marinoan glaciation. This may indicate a buildup of carbon dioxide in the atmosphere, some of which would dissolve into the oceans to form carbonic acid. Although the boron variations may be evidence of extreme climate change, they need not imply a global glaciation.

Space Dust

Earth's surface is very depleted in the element iridium, which primarily resides in the Earth's core. The only significant source of the element at the surface is cosmic particles that reach Earth. During a snowball Earth, iridium would accumulate on the ice sheets, and when the ice melted the resulting layer of sediment would be rich in iridium. An iridium anomaly has been discovered at the base of the cap carbonate formations, and has been used to suggest that the glacial episode lasted for at least 3 million years, but this does not necessarily imply a *global* extent to the glaciation; indeed, a similar anomaly could be explained by the impact of a large meteorite.

Cyclic Climate Fluctuations

Using the ratio of mobile cations to those that remain in soils during chemical weathering (the chemical index of alteration), it has been shown that chemical weathering varied in a cyclic fashion within a glacial succession, increasing during interglacial periods and decreasing during cold and arid glacial periods. This pattern, if a true reflection of events, suggests that the "snowball Earths" bore a stronger resemblance to Pleistocene ice age cycles than to a completely frozen Earth.

What's more, glacial sediments of the Portaskaig formation in Scotland clearly show interbedded cycles of glacial and shallow marine sediments. The significance of these deposits is highly reliant upon their dating. Glacial sediments are difficult to date, and the closest dated bed to the Portaskaig group is 8 km stratigraphically above the beds of interest. Its dating to 600 Ma means the beds can be tentatively correlated to the Sturtian glaciation, but they may represent the advance or retreat of a snowball Earth.

Mechanisms

The initiation of a snowball Earth event would involve some initial cooling mechanism, which would result in an increase in Earth's coverage of snow and ice. The increase in Earth's coverage of snow and ice would in turn increase Earth's albedo, which would result in positive feedback for cooling. If enough snow and ice accumulates, run-away cooling would result. This positive feedback is facilitated by an equatorial continental distribution, which would allow ice to accumulate in the regions closer to the equator, where solar radiation is most direct.

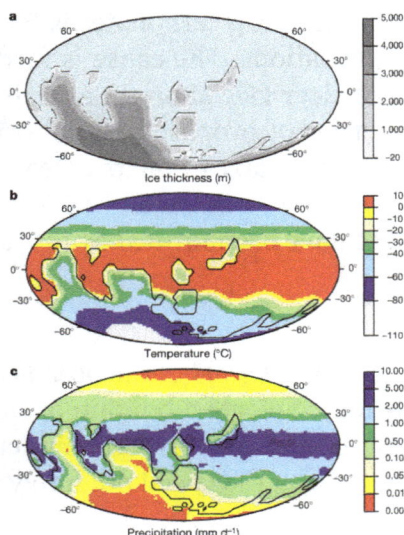

One computer simulation of conditions during a snowball Earth period

Many possible triggering mechanisms could account for the beginning of a snowball Earth, such as the eruption of a supervolcano, a reduction in the atmospheric concentration of greenhouse gases such as methane and/or carbon dioxide, changes in Solar energy output, or perturbations of Earth's orbit. Regardless of the trigger, initial cooling results in an increase in the area of Earth's surface covered by ice and snow, and the additional ice and snow reflects more Solar energy back to space, further cooling Earth and further increasing the area of Earth's surface covered by ice and snow. This positive feedback loop could eventually produce a frozen equator as cold as modern Antarctica.

Global warming associated with large accumulations of carbon dioxide in the atmosphere over millions of years, emitted primarily by volcanic activity, is the proposed trigger for melting a snowball Earth. Due to positive feedback for melting, the eventual melting of the snow and ice covering most of Earth's surface would require as little as a millennium.

Continental Distribution

A tropical distribution of the continents is, perhaps counter-intuitively, necessary to allow the initiation of a snowball Earth. Firstly, tropical continents are more reflective than open ocean, and so absorb less of the Sun's heat: most absorption of Solar energy on Earth today occurs in tropical oceans.

Further, tropical continents are subject to more rainfall, which leads to increased river discharge — and erosion. When exposed to air, silicate rocks undergo weathering reactions which remove carbon dioxide from the atmosphere. These reactions proceed in the general form: Rock-forming mineral + CO_2 + H_2O → cations + bicarbonate + SiO_2. An example of such a reaction is the weathering of wollastonite:

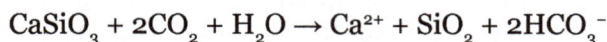

$$CaSiO_3 + 2CO_2 + H_2O \rightarrow Ca^{2+} + SiO_2 + 2HCO_3^-$$

The released calcium cations react with the dissolved bicarbonate in the ocean to form calcium carbonate as a chemically precipitated sedimentary rock. This transfers carbon dioxide, a greenhouse gas, from the air into the geosphere, and, in steady-state on geologic time scales, offsets the carbon dioxide emitted from volcanoes into the atmosphere.

A paucity of suitable sediments for analysis makes precise continental distribution during the Neoproterozoic difficult to establish. Some reconstructions point towards polar continents — which have been a feature of all other major glaciations, providing a point upon which ice can nucleate. Changes in ocean circulation patterns may then have provided the trigger of snowball Earth.

Additional factors that may have contributed to the onset of the Neoproterozoic snowball include the introduction of atmospheric free oxygen, which may have reached sufficient quantities to react with methane in the atmosphere, oxidizing it to carbon dioxide, a much weaker greenhouse gas, and a younger — thus fainter — Sun, which would have emitted 6 percent less radiation in the Neoproterozoic.

Normally, as Earth gets colder due to natural climatic fluctuations and changes in incoming solar radiation, the cooling slows these weathering reactions. As a result, less carbon dioxide is removed from the atmosphere and Earth warms as this greenhouse gas accumulates — this 'negative feedback' process limits the magnitude of cooling. During the Cryogenian period, however, Earth's continents were all at tropical latitudes, which made this moderating process less effective, as high weathering rates continued on land even as Earth cooled. This let ice advance beyond the polar regions. Once ice advanced to within 30° of the equator, a positive feedback could ensue such that the increased reflectiveness (albedo) of the ice led to further cooling and the formation of more ice, until the whole Earth is ice-covered.

Polar continents, due to low rates of evaporation, are too dry to allow substantial carbon deposition — restricting the amount of atmospheric carbon dioxide that can be removed from the carbon cycle. A gradual rise of the proportion of the isotope carbon-13 relative to carbon-12 in sediments pre-dating "global" glaciation indicates that CO_2 draw-down before snowball Earths was a slow and continuous process.

The start of snowball Earths are always marked by a sharp downturn in the $\delta^{13}C$ value of sediments, a hallmark that may be attributed to a crash in biological productivity as a result of the cold temperatures and ice-covered oceans.

In January 2016, Gernon et al. proposed a "shallow-ridge hypothesis" involving the breakup of the supercontinent Rodinia, linking the eruption and rapid alteration of hyaloclastites along shallow ridges to massive increases in alkalinity in an ocean with thick ice cover. Gernon et al. demonstrated that the increase in alkalinity over the course of glaciation is sufficient to explain the thickness of cap carbonates formed in the aftermath of Snowball Earth events.

During the Frozen Period

Global ice sheets may have created the bottleneck required for the evolution of multicellular life.

Global temperature fell so low that the equator was as cold as modern-day Antarctica. This low temperature was maintained by the high albedo of the ice sheets, which reflected most incoming solar energy into space. A lack of heat-retaining clouds, caused by water vapor freezing out of the atmosphere, amplified this effect.

Breaking Out of Global Glaciation

The carbon dioxide levels necessary to unfreeze Earth have been estimated as being 350 times what they are today, about 13% of the atmosphere. Since the Earth was almost completely covered with ice, carbon dioxide could not be withdrawn from the atmosphere by release of alkaline metal ions weathering out of siliceous rocks. Over 4 to 30 million years, enough CO_2 and methane, mainly emitted by volcanoes, would accumulate to finally cause enough greenhouse effect to make surface ice melt in the tropics until a band of permanently ice-free land and water developed; this would be darker than the ice, and thus absorb more energy from the Sun — initiating a "positive feedback".

Destabilization of substantial deposits of methane hydrates locked up in low-latitude permafrost may also have acted as a trigger and/or strong positive feedback for deglaciation and warming.

On the continents, the melting of glaciers would release massive amounts of glacial deposit, which would erode and weather. The resulting sediments supplied to the ocean would be high in nutrients such as phosphorus, which combined with the abundance of CO_2 would trigger a cyanobacteria population explosion, which would cause a relatively rapid reoxygenation of the atmosphere, which may have contributed to the rise of the Ediacaran biota and the subsequent Cambrian explosion — a higher oxygen concentration allowing large multicellular lifeforms to develop. Although the positive feedback loop would melt the ice in geological short order, perhaps less than 1,000 years, replenishment of atmospheric oxygen and depletion of the CO_2 levels would take further millennia.

It is possible that carbon dioxide levels fell enough for Earth to freeze again; this cycle may have repeated until the continents had drifted to more polar latitudes.

More recent evidence suggests that with colder oceanic temperatures, the resulting higher ability of the oceans to dissolve gases led to the carbon content of sea water being more quickly oxidized to carbon dioxide. This leads directly to an increase of atmospheric carbon dioxide, enhanced greenhouse warming of Earth's surface, and the prevention of a total snowball state.

Slushball Earth Hypothesis

While the presence of glaciers is not disputed, the idea that the entire planet was covered in ice is more contentious, leading some scientists to posit a "slushball Earth", in which a band of ice-free, or ice-thin, waters remains around the equator, allowing for a continued hydrologic cycle.

This hypothesis appeals to scientists who observe certain features of the sedimentary record that can only be formed under open water, or rapidly moving ice (which would require somewhere ice-free to move to). Recent research observed geochemical cyclicity in clastic rocks, showing that the "snowball" periods were punctuated by warm spells, similar to ice age cycles in recent Earth history. Attempts to construct computer models of a snowball Earth have also struggled to accommodate global ice cover without fundamental changes in the laws and constants which govern the planet.

A less extreme snowball Earth hypothesis involves continually evolving continental configurations and changes in ocean circulation. Synthesised evidence has produced models indicating a "slushball Earth", where the stratigraphic record does not permit postulating complete global glaciations. Kirschivink's original hypothesis had recognised that warm tropical puddles would be expected to exist in a snowball earth.

The snowball Earth hypothesis does not explain the alternation of glacial and interglacial events, nor the oscillation of glacial sheet margins.

Scientific Dispute

The argument against the hypothesis is evidence of fluctuation in ice cover and melting during "snowball Earth" deposits. Evidence for such melting comes from evidence of glacial dropstones, geochemical evidence of climate cyclicity, and interbedded glacial and shallow marine sediments. A longer record from Oman, constrained to 13°N, covers the period from 712 to 545 million years ago — a time span containing the Sturtian and Marinoan glaciations — and shows both glacial and ice-free deposition.

There have been difficulties in recreating a snowball Earth with global climate models. Simple GCMs with mixed-layer oceans can be made to freeze to the equator; a more sophisticated model with a full dynamic ocean (though only a primitive sea ice model) failed to form sea ice to the equator. In addition, the levels of CO_2 necessary to melt a global ice cover have been calculated to be 130,000 ppm, which is considered by some to be unreasonably large.

Strontium isotopic data have been found to be at odds with proposed snowball Earth models of silicate weathering shutdown during glaciation and rapid rates immediately post-glaciation. Therefore, methane release from permafrost during marine transgression was proposed to be the source of the large measured carbon excursion in the time immediately after glaciation.

"Zipper Rift" Hypothesis

Nick Eyles suggest that the Neoproterozoic Snowball Earth was in fact no different from any other glaciation in Earth's history, and that efforts to find a single cause are likely to end in failure. The "Zipper rift" hypothesis proposes two pulses of continental "unzipping" — first, the breakup of the supercontinent Rodinia, forming the proto-Pacific ocean; then the splitting of the continent Baltica from Laurentia, forming the proto-Atlantic — coincided with the glaciated periods. The associated tectonic uplift would form high plateaus, just as the East African Rift is responsible for high topography; this high ground could then host glaciers.

Banded iron formations have been taken as unavoidable evidence for global ice cover, since they require dissolved iron ions and anoxic waters to form; however, the limited extent of the Neoproterozoic banded iron deposits means that they may not have formed in frozen oceans, but instead in inland seas. Such seas can experience a wide range of chemistries; high rates of evaporation could concentrate iron ions, and a periodic lack of circulation could allow anoxic bottom water to form.

Continental rifting, with associated subsidence, tends to produce such landlocked water bodies. This rifting, and associated subsidence, would produce the space for the fast deposition of sediments, negating the need for an immense and rapid melting to raise the global sea levels.

High-obliquity Hypothesis

A competing hypothesis to explain the presence of ice on the equatorial continents was that Earth's axial tilt was quite high, in the vicinity of 60°, which would place Earth's land in high "latitudes", although supporting evidence is scarce. A less extreme possibility would be that it was merely Earth's magnetic pole that wandered to this inclination, as the magnetic readings which suggested ice-filled continents depends on the magnetic and rotational poles being relatively similar. In either of these two situations, the freeze would be limited to relatively small areas, as is the case today; severe changes to Earth's climate are not necessary.

Inertial Interchange True Polar Wander

The evidence for low-latitude glacial deposits during the supposed snowball Earth episodes has been reinterpreted via the concept of inertial interchange true polar wander (IITPW). This hypothesis, created to explain palaeomagnetic data, suggests that Earth's axis of rotation shifted one or more times during the general time-frame attributed to snowball Earth. This could feasibly produce the same distribution of glacial deposits without requiring any of them to have been deposited at equatorial latitude. While the physics behind the proposition is sound, the removal of one flawed data point from the original study rendered the application of the concept in these circumstances unwarranted.

Several alternative explanations for the evidence have been proposed.

Survival of Life Through Frozen Periods

A tremendous glaciation would curtail photosynthetic life on Earth, thus letting the atmospheric oxygen be drastically depleted and perhaps even disappear, and thus allow non-oxidized iron-rich rocks to form.

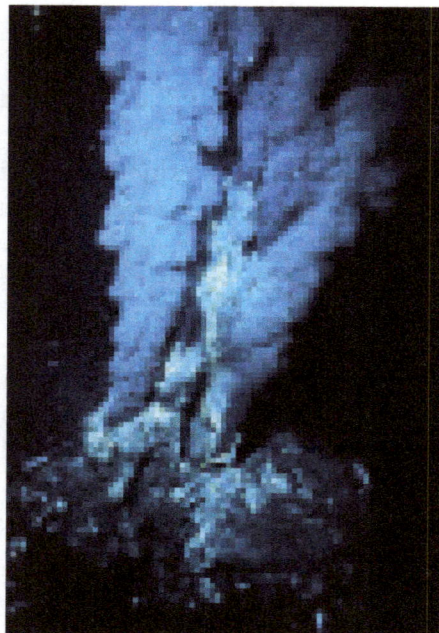

A black smoker, a type of hydrothermal vent

Detractors argue that this kind of glaciation would have made life extinct entirely. However, micro-fossils such as stromatolites and oncolites prove that, in shallow marine environments at least, life did not suffer any perturbation. Instead life developed a trophic complexity and survived the cold period unscathed. Proponents counter that it may have been possible for life to survive in these ways:

- In reservoirs of anaerobic and low-oxygen life powered by chemicals in deep oceanic hy-drothermal vents surviving in Earth's deep oceans and crust; but photosynthesis would not have been possible there.

- As eggs and dormant cells and spores deep-frozen into ice during the most severe phases of the frozen period.

- Under the ice layer, in chemolithotrophic (mineral-metabolizing) ecosystems theoretically re-sembling those in existence in modern glacier beds, high-alpine and Arctic talus permafrost, and basal glacial ice. This is especially plausible in areas of volcanism or geothermal activity.

- In deep ocean regions far from the supercontinent Rodinia or its remnants as it broke apart and drifted on the tectonic plates, which may have allowed for some small regions of open water preserving small quantities of life with access to light and CO_2 for photosynthesizers (not multicellular plants, which did not yet exist) to generate traces of oxygen that were enough to sustain some oxygen-dependent organisms. This would happen even if the sea froze over completely, if small parts of the ice were thin enough to admit light.

- In nunatak areas in the tropics, where daytime tropical sun or volcanic heat heated bare rock sheltered from cold wind and made small temporary melt pools, which would freeze at sunset.

- In pockets of liquid water within and under the ice caps, similar to Lake Vostok in Antarc-tica. In theory, this system may resemble microbial communities living in the perennially frozen lakes of the Antarctic dry valleys. Photosynthesis can occur under ice up to 100 m thick, and at the temperatures predicted by models equatorial sublimation would prevent equatorial ice thickness from exceeding 10 m.

- In small oases of liquid water, as would be found near geothermal hotspots resembling Iceland today.

However, organisms and ecosystems, as far as it can be determined by the fossil record, do not ap-pear to have undergone the significant change that would be expected by a mass extinction. With the advent of more precise dating, a phytoplankton extinction event which had been associated with snowball Earth was shown to precede glaciations by 16 million years. Even if life were to cling on in all the ecological refuges listed above, a whole-Earth glaciation would result in a biota with a noticeably different diversity and composition. This change in diversity and composition has not yet been observed– in fact, the organisms which should be most susceptible to climatic variation emerge unscathed from the snowball Earth.

Implications

A snowball Earth has profound implications in the history of life on Earth. While many refugia have been postulated, global ice cover would certainly have ravaged ecosystems dependent on sunlight. Geochemical evidence from rocks associated with low-latitude glacial deposits have been interpreted to show a crash in oceanic life during the glacials.

The melting of the ice may have presented many new opportunities for diversification, and may indeed have driven the rapid evolution which took place at the end of the Cryogenian period.

Effect on Early Evolution

The Neoproterozoic was a time of remarkable diversification of multicellular organisms, including animals. Organism size and complexity increased considerably after the end of the snowball glaciations. This development of multicellular organisms may have been the result of increased evolutionary pressures resulting from multiple icehouse-hothouse cycles; in this sense, snowball Earth episodes may have "pumped" evolution. Alternatively, fluctuating nutrient levels and rising oxygen may have played a part. Interestingly, another major glacial episode may have ended just a few million years before the Cambrian explosion.

Dickinsonia costata, an Ediacaran organism of unknown affinity, with a quilted appearance

Mechanistically, the effect of snowball Earth (in particular the later glaciations) on complex life is likely to have occurred through the process of kin selection. Organ-scale differentiation, in particular the terminal (irreversible) differentiation present in animals, requires the individual cell (and the genes contained within it) to "sacrifice" their ability to reproduce, so that the colony is not disrupted. From the short-term perspective of the gene, more offspring will be gained by causing the cell in which it is contained to ignore any signals received from the colony, and to reproduce at the maximum rate, regardless of the implications for the wider group. Today, this incentive explains the formation of tumours in animals and plants.

It has been argued that because snowball Earth would undoubtedly have decimated the population size of any given species, the extremely small populations that resulted would all have been descended from a small number of individuals, and consequently the average relatedness between any two individuals (in this case individual cells) would have been exceptionally high as a result of glaciations. Altruism is known to increase from rarity when relatedness (R) exceeds the ratio of the cost (C) to the altruist (in this case, the cell giving up its own reproduction by differentiating), to the benefit (B) to the recipient of altruism (the germ line of the colony, that reproduces as a result of the differentiation), i.e. $R > C/B$. The evolutionary pressure of the high relatedness in the context of a post-glaciation population boom may have been sufficient to overcome the reproductive cost of forming a complex animal, for the first time in Earth's history.

There is also a rival hypothesis which has been gaining currency in recent years: that early snowball Earths did not so much *affect* the evolution of life on Earth as result from it. In fact the two hypotheses are not mutually exclusive. The idea is that Earth's life forms affect the global carbon cycle and so major evolutionary events alter the carbon cycle, redistributing carbon within various reservoirs within the biosphere system and in the process temporarily lowering the atmospheric (greenhouse) carbon reservoir until the revised biosphere system settled into a new state. The Snowball I episode (of the Huronian glaciation 2.4 to 2.1 billion years) and Snowball II (of the Precambrian's Cryogenian between 580 – 850 million years and which itself had a number of distinct episodes) are respectively thought to be caused by the evolution of oxygenic photosynthesis and then the rise of more advanced multicellular animal life and life's colonization of the land.

Effects on Ocean Circulation

Global ice cover, if it existed, may – in concert with geothermal heating – have led to a lively, well mixed ocean with great vertical convective circulation.

Occurrence and Timing of Snowball Earths

Neoproterozoic

There are three or four significant ice ages during the late Neoproterozoic. Of these, the Marinoan was the most significant, and the Sturtian glaciations were also truly widespread. Even the leading snowball proponent Hoffman agrees that the ~million-year-long Gaskiers glaciation did not lead to global glaciation, although it was probably as intense as the late Ordovician glaciation. The status of the Kaigas "glaciation" or "cooling event" is currently unclear; some workers do not recognise it as a glacial, others suspect that it may reflect poorly dated strata of Sturtian association, and others believe it may indeed be a third ice age. It was certainly less significant than the Sturtian or Marinoan glaciations, and probably not global in extent. Emerging evidence suggests that the Earth underwent a number of glaciations during the Neoproterozoic, which would stand strongly at odds with the snowball hypothesis.

Paleoproterozoic

The snowball Earth hypothesis has been invoked to explain glacial deposits in the Huronian Supergroup of Canada, though the palaeomagnetic evidence that suggests ice sheets at low latitudes is contested. The glacial sediments of the Makganyene formation of South Africa are slightly younger than the Huronian glacial deposits (~2.25 billion years old) and were deposited at tropical latitudes. It has been proposed that rise of free oxygen that occurred during the Great Oxygenation Event removed methane in the atmosphere through oxidation. As the Sun was notably weaker at the time, Earth's climate may have relied on methane, a powerful greenhouse gas, to maintain surface temperatures above freezing.

In the absence of this methane greenhouse, temperatures plunged and a snowball event could have occurred.

Karoo Ice Age

Before the theory of continental drift, glacial deposits in Carboniferous strata in tropical continents areas such as India and South America led to speculation that the Karoo Ice Age glaciation reached into the tropics. However, a continental reconstruction shows that ice was in fact constrained to the polar parts of the supercontinent Gondwana.

Endemism

Endemism is the ecological state of a species being unique to a defined geographic location, such as an island, nation, country or other defined zone, or habitat type; organisms that are indigenous to a place are not endemic to it if they are also found elsewhere. The extreme opposite of endemism is cosmopolitan distribution. An alternative term for a species that is endemic is precinctive, which applies to species (and subspecific categories) that are restricted to a defined geographical area.

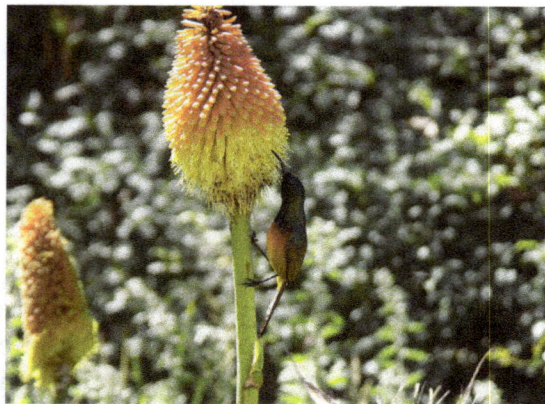

The orange-breasted sunbird (*Nectarinia violacea*) is exclusively found in fynbos vegetation.

Bicolored frog (*Clinotarsus curtipes*) is endemic to the Western Ghats of India

The word *endemic* is from New Latin *endēmicus*, *endēmos*, "native." *Endēmos* is formed of *en* meaning "in," and *dēmos* meaning "the people." The term, *precinctive*, has been suggested by some scientists,[a] and was first used in botany by MacCaughey in 1917. It is the equivalent of "endemism". *Precinction* was perhaps first used by Frank and McCoy. *Precinctive* seems to have been coined by David Sharp of the Hawaiian fauna in 1900: "I use the word precinc-tive in the sense of 'confined to the area under discussion' ... 'precinctive forms' means those forms that are confined to the area specified." That definition excludes artificial confinement of examples by humans in far-off botanical gardens or zoological parks.

Overview

Physical, climatic, and biological factors can contribute to endemism. The orange-breasted sun-bird is exclusively found in the fynbos vegetation zone of southwestern South Africa. The glacier bear is found only in limited places in Southeast Alaska. Political factors can play a part if a species is protected, or actively hunted, in one jurisdiction but not another.

There are two subcategories of endemism: paleoendemism and neoendemism. Paleoendemism refers to species that were formerly widespread but are now restricted to a smaller area. Neoendemism refers to species that have recently arisen, such as through divergence and reproductive isolation or through hybridization and polyploidy in plants.

Endemic types or species are especially likely to develop on geographically and biologically isolated areas such as islands and remote island groups, such as Hawaii, the Galápagos Islands, and Socotra; they can equally develop in biologically isolated areas such as the highlands of Ethiopia, or large bodies of water far from other lakes, like Lake Baikal.

Endemics can easily become endangered or extinct if their restricted habitat changes, particularly—but not only—due to human actions, including the introduction of new organisms. There were millions of both Bermuda petrels and "Bermuda cedars" (actually *junipers*) in Bermuda when it was settled at the start of the seventeenth century. By the end of the century, the petrels were thought extinct. Cedars, already ravaged by centuries of shipbuilding, were driven nearly to extinction in the twentieth century by the introduction of a parasite. Bermuda petrels and cedars are now rare, as are other species endemic to Bermuda.

Threats to Highly Endemistic Regions

Principal causes of habitat degradation and loss in highly endemistic ecosystems include agriculture, urban growth, surface mining, mineral extraction, logging operations and slash-and-burn agriculture.

Biodiversity Hotspot

A biodiversity hotspot is a biogeographic region with significant levels of biodiversity that is under threat from humans. Norman Myers wrote about the concept in two articles in "The Environmentalist" (1988), & 1990 revised after thorough analysis by Myers and others in "Hotspots: Earth's Biologically Richest and Most Endangered Terrestrial Ecoregions" and a paper published in the journal *Nature*

To qualify as a biodiversity hotspot on Myers 2000 edition of the hotspot-map, a region must meet two strict criteria: it must contain at least 0.5% or 1,500 species of vascular plants as endemics, and it has to have lost at least 70% of its primary vegetation. Around the world, 34 areas qualify under this definition, with nine other possible candidates. These sites support nearly 60% of the world's plant, bird, mammal, reptile, and amphibian species, with a very high share of those species as endemics.

Hotspot Conservation Initiatives

Only a small percentage of the total land area within biodiversity hotspots is now protected. Several international organizations are working in many ways to conserve biodiversity hotspots.

- Critical Ecosystem Partnership Fund (CEPF) is a global program that provides funding and technical assistance to nongovernmental organizations and participation to protect the Earth's richest regions of plant and animal diversity including: biodiversity hotspots, high-biodiversity wilderness areas and important marine regions.

- The World Wide Fund for Nature has derived a system called the "Global 200 Ecoregions", the aim of which is to select priority Ecoregions for conservation within each of 14 terrestrial, 3 freshwater, and 4 marine habitat types. They are chosen for their species richness, endemism, taxonomic uniqueness, unusual ecological or evolutionary phenomena, and global rarity. All biodiversity hotspots contain at least one Global 200 Ecoregion.

- Birdlife International has identified 218 "Endemic Bird Areas" (EBAs) each of which hold two or more bird species found nowhere else. Birdlife International has identified more than 11,000 Important Bird Areas all over the world.

- Plant life International coordinates several the world aiming to identify Important Plant Areas.

- Alliance for Zero Extinction is an initiative of a large number of scientific organizations and conservation groups who co-operate to focus on the most threatened endemic species of the world. They have identified 595 sites, including a large number of Birdlife's Important Bird Areas.

- The National Geographic Society has prepared a world map of the hotspots and ArcView shapefile and metadata for the Biodiversity Hotspots including details of the individual endangered fauna in each hotspot, which is available from Conservation International.

Distribution by Region

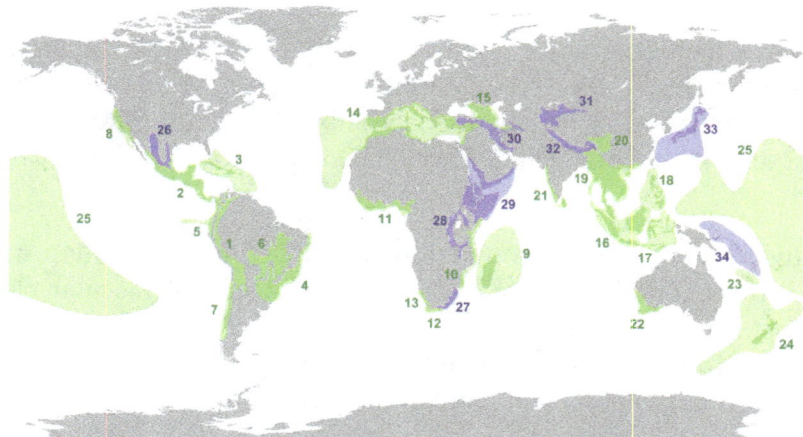

Biodiversity hotspots. Original proposal in green, and added regions in blue.

North and Central America

- California Floristic Province •8•
- Madrean pine-oak woodlands •26•
- Mesoamerica •2•

The Caribbean

- Caribbean Islands •3•

South America

- Atlantic Forest •4•
- Cerrado •6•
- Chilean Winter Rainfall-Valdivian Forests •7•
- Tumbes-Chocó-Magdalena •5•
- Tropical Andes •1•

Europe

- Mediteranean Basin •14•

Africa

- Cape Floristic Region •12•
- Coastal Forests of Eastern Africa •10•
- Eastern Afromontane •28•
- Guinean Forests of West Africa •11•
- Horn of Africa •29•
- Madagascar and the Indian Ocean Islands •9•
- Maputaland-Pondoland-Albany •27•
- Succulent Karoo •13•

Central Asia

- Mountains of Central Asia •31•

South Asia

- Eastern Himalaya, Nepal •32•
- Indo-Burma, India and Myanmar •19•

- Western Ghats, India•21•
- Sri Lanka, Sri Lanka•21•

South East Asia and Asia-Pacific

- East Melanesian Islands •34•
- New Caledonia •23•
- New Zealand •24•
- Philippines •18•
- Polynesia-Micronesia •25•
- Southwest Australia •22•
- Sundaland •16•
- Wallacea •17•

East Asia

- Japan •33•
- Mountains of Southwest China •20•

West Asia

- Caucasus •15•
- Irano-Anatolian •30•

Bold text Critiques of "Hotspots" The high profile of the biodiversity hotspots approach has resulted in some criticism. Papers such as Kareiva & Marvier (2003) have argued that the biodiversity hotspots:

- Do not adequately represent other forms of species richness (e.g. total species richness or threatened species richness).
- Do not adequately represent taxa other than vascular plants (e.g. vertebrates, or fungi).
- Do not protect smaller scale richness hotspots.
- Do not make allowances for changing land use patterns. Hotspots represent regions that have experienced considerable habitat loss, but this does not mean they are experiencing ongoing habitat loss. On the other hand, regions that are relatively intact (e.g. the Amazon Basin) have experienced relatively little land loss, but are currently losing habitat at tremendous rates.
- Do not protect ecosystem services.
- Do not consider phylogenetic diversity.

A recent series of papers has pointed out that biodiversity hotspots (and many other priority region sets) do not address the concept of cost. The purpose of biodiversity hotspots is not simply to identify regions that are of high biodiversity value, but to prioritize conservation spending. The regions identified include some in the developed world (e.g. the California Floristic Province), alongside others in the developing world (e.g. Madagascar). The cost of land is likely to vary between these regions by an order of magnitude or more, but the biodiversity hotspot designations do not consider the conservation importance of this difference. However, the available resources for conservation also tend to vary in this way.

References

- Schoener, Thomas W. (2009). "Ecological Niche". In Simon A. Levin. The Princeton Guide to Ecology. Princeton: Princeton University Press. pp. 2–13. ISBN 978-0-691-12839-9.

- Lindenmayer, David B.; Gene E. Likens (2010). "The Problematic, the Effective and the Ugly – Some Case Studies". Effective Ecological Monitoring. Collingwood, Australia: CSIRO Publishing. pp. 87–145. ISBN 978-1-84971-145-6.

- Kunin, W.E.; Gaston, Kevin, eds. (31 December 1996). The Biology of Rarity: Causes and consequences of rare—common differences. ISBN 978-0412633805. Retrieved 26 May 2015.

- Stearns, Beverly Peterson; Stearns, S. C.; Stearns, Stephen C. (2000). Watching, from the Edge of Extinction. Yale University Press. p. 1921. ISBN 978-0-300-08469-6. Retrieved 2014-12-27.

- Wilson, E.O., The Future of Life (2002) (ISBN 0-679-76811-4). Leakey, Richard, The Sixth Extinction: Patterns of Life and the Future of Humankind, ISBN 0-385-46809-1

- Davis, Paul and Kenrick, Paul. Fossil Plants. Smithsonian Books, Washington D.C. (2004). Morran, Robin, C.; A Natural History of Ferns. Timber Press (2004). ISBN 0-88192-667-1

- Niles Eldredge, Time Frames: Rethinking of Darwinian Evolution and the Theory of Punctuated Equilibria, 1986, Heinemann ISBN 0-434-22610-6

- Mills, L. Scott (2009-03-12). Conservation of Wildlife Populations: Demography, Genetics and Management. John Wiley & Sons. p. 13. ISBN 9781444308938.

- Stearns, Beverly Peterson and Stephen C. (2000). "Preface". Watching, from the Edge of Extinction. Yale University Press. pp. x. ISBN 0-300-08469-2.

- Clover, Charles (2004). The End of the Line: How overfishing is changing the world and what we eat. London: Ebury Press. ISBN 0-09-189780-7.

- Benton M J (2005). When Life Nearly Died: The Greatest Mass Extinction of All Time. Thames & Hudson. ISBN 978-0-500-28573-2.

- Primack, Richard (2014). Essentials of Conservation Biology. Sunderland, MA USA: Sinauer Associates, Inc. Publishers. pp. 217–245. ISBN 9781605352893.

- G. Miller; Scott Spoolman (2012). Environmental Science – Biodiversity Is a Crucial Part of the Earth's Natural Capital. Cengage Learning. p. 62. ISBN 1-133-70787-4. Retrieved 2014-12-27.

3

Different Types of Species

Species can be classified according to their geological availability and the threat that humans pose to them. The chapter discusses the various types of species classification like rare, threatened, extinct, keystone, indicator, umbrella, flagship, introduced and invasive species. This chapter emphasizes about the need for preservation and conservation of species while also discussing the role that different species play in maintaining ecological balance.

Rare Species

A rare species is a group of organisms that are very uncommon, scarce, or infrequently encountered. This designation may be applied to either a plant or animal taxon, and may be distinct from the term *endangered* or *threatened species*. Designation of a rare species may be made by an official body, such as a national government, state, or province. However, the term more commonly appears without reference to specific criteria. The IUCN does not normally make such designations, but may use the term in scientific discussion.

Rarity rests on a specific species being represented by a small number of organisms worldwide, usually fewer than 10,000. However, a species having a very narrow endemic range or fragmented habitat also influences the concept. Rare species are not uncommon, since nearly 75% of known species are rare.

A species may be endangered or vulnerable, but not considered rare if—for example—it has a large, dispersed population, but its numbers are declining rapidly or predicted to do so. Rare species are generally considered threatened because a small population size is more likely to not recover from stochastic events (things that *could* happen).

Rare species are species with small populations. Many move into the endangered or vulnerable category if the negative factors affecting them continue to operate. Examples of rare species are the Himalayan brown deer, desert fox, wild Asiatic buffalo and hornbill.

A rare plant's legal status can be observed through the USDA's Plants Database.

Common name	Scientific name	Conservation status	Population	Global range
Giant panda	*Ailuropoda melanoleuca*	Endangered	1,000 to 3,000	China
Wild Bactrian camel	*Camelus ferus*	Critically endangered	950	Kazakhstan/Northwest China/ Southern Mongolia
Cheetah	*Acinonyx jubatus*	Vulnerable	7,000 to 10,000	Africa/Southwestern Asia
California condor	*Gymnogyps californianus*	Critically endangered	130 approx.	West North America

Alagoas curassow	*Mitu mitu*	Extinct in the wild	130 (in captivity)	North East Brazil
Philippine eagle	*Pithecophaga jefferyi*	Critically endangered	200 breeding pairs	Eastern Luzon, Samar, Leyte, and Mindanao
Black softshell turtle	*Nilssonia nigricans*	Extinct in the wild	150 to 300 (in captivity)	Hazrat Sultan Bayazid Bastami shrine at Chittagong
Key-tree cactus	*Pilosocereus robinii*	Endangered	7 to 15	Florida Keys , Mexico , Puerto Rico

Endangered Species

An Endangered (EN) species is a species which has been categorized by the International Union for Conservation of Nature (IUCN) Red List as likely to become extinct. "Endangered" is the second most severe conservation status for wild populations in the IUCN's schema after Critically Endangered (CR).

In 2012, the IUCN Red List featured 3079 animal and 2655 plant species as endangered (EN) worldwide. The figures for 1998 were, respectively, 1102 and 1197

Many nations have laws that protect conservation-reliant species: for example, forbidding hunting, restricting land development or creating preserves. Population numbers, trends and species' conservation status can be found in the lists of organisms by population.

Conservation Status

The conservation status of a species indicates the likelihood that it will become extinct. Many factors are considered when assessing the conservation status of a species; e.g., such statistics as the number remaining, the overall increase or decrease in the population over time, breeding success rates, or known threats. The IUCN Red List of Threatened Species is the best-known worldwide conservation status listing and ranking system.

Over 40% of the world's species are estimated to be at risk of extinction. Internationally, 199 countries have signed an accord to create Biodiversity Action Plans that will protect endangered and other threatened species. In the United States, such plans are usually called Species Recovery Plans.

IUCN Red List

The Siberian tiger is an Endangered (EN) tiger subspecies. Three tiger subspecies are already extinct

Blue-throated macaw, an endangered species

Siamese crocodile, an endangered species

Kemp's ridley sea turtle, an endangered species

Though labelled a list, the IUCN Red List is a system of assessing the global conservation status of species that includes "Data Deficient" (DD) species – species for which more data and assessment is required before their status may be determined – as well species comprehensively assessed by the IUCN's species assessment process. Those species of "Near Threatened" (NT) and "Least Concern" (LC) status have been assessed and found to have relatively robust and healthy populations, though these may be in decline. Unlike their more general use elsewhere, the List uses the terms "endangered species" and "threatened species" with particular meanings: "Endangered" (EN) species lie between "Vulnerable" (VU) and "Critically Endangered" (CR) species, while "Threatened" species are those species determined to be Vulnerable, Endangered or Critically Endangered.

The IUCN categories, with examples of animals classified by them, include:

Extinct (EX)

- Examples: aurochs

- Bali tiger
- blackfin cisco
- Caribbean monk seal
- Carolina parakeet
- Caspian tiger
- dodo
- dusky seaside sparrow
- eastern cougar
- golden toad
- great auk
- Japanese sea lion
- Javan tiger
- Labrador duck
- passenger pigeon
- Schomburgk's deer
- Steller's sea cow
- thylacine
- toolache wallaby
- western black rhinoceros
- California Grizzly Bear

Extinct in the wild (EW)

Captive individuals survive, but there is no free-living, natural population.

- Examples: Barbary lion
- Hawaiian crow
- Père David's deer
- scimitar oryx
- Socorro dove
- Wyoming toad

Critically endangered (CR)

Faces an extremely high risk of extinction in the immediate future.

- Examples: addax
- African wild ass
- Alabama cavefish
- Amur leopard
- Arakan forest turtle
- Asiatic cheetah
- axolotl
- Bactrian camel
- black rhino
- blue-throated macaw
- Brazilian merganser
- brown spider monkey
- California condor
- Chinese alligator
- Chinese giant salamander
- gharial
- Hawaiian monk seal
- Javan rhino
- kakapo
- Leadbeater's possum
- Mediterranean monk seal
- mountain gorilla
- northern hairy-nosed wombat
- Philippine eagle
- red wolf
- saiga
- Siamese crocodile
- Malayan tiger
- Spix's macaw
- southern bluefin tuna

- South China tiger
- Sumatran orangutan
- Sumatran rhinoceros
- Sumatran tiger
- vaquita
- Yangtze river dolphin
- northern white rhinoceros
- hawksbill sea turtle
- Kemp's ridley sea turtle

Endangered (EN)

Faces a high risk of extinction in the near future.

- Examples: African penguin
- African wild dog
- Asian elephant
- Asiatic lion
- Australasian bittern
- blue whale
- bonobo
- Bornean orangutan
- common chimpanzee
- dhole
- eastern lowland gorilla
- hispid hare
- giant otter
- giant panda
- Goliath frog
- green sea turtle
- loggerhead sea turtle
- Grevy's zebra
- hyacinth macaw

- Humblot's heron
- Iberian lynx
- Japanese crane
- Japanese night heron
- Lear's macaw
- Malayan tapir
- markhor
- Malagasy pond heron
- Persian leopard
- proboscis monkey
- purple-faced langur
- pygmy hippopotamus
- red-breasted goose
- Rothschild's giraffe
- snow leopard
- South Andean deer
- Sri Lankan elephant
- takhi (near Critically Endangered) Toque macaque
- Vietnamese pheasant
- volcano rabbit
- wild water buffalo
- white-eared night heron
- fishing cat
- tasmanian devil

Vulnerable (VU)

Faces a high risk of endangerment in the medium term.

- Examples: African grey parrot
- African bush elephant
- African lion
- American paddlefish

- common carp
- clouded leopard
- cheetah
- dugong
- Far Eastern curlew
- fossa
- Galapagos tortoise
- gaur
- blue-eyed cockatoo
- golden hamster
- whale shark
- hippopotamus
- Humboldt penguin
- Indian rhinoceros
- Komodo dragon
- lesser white-fronted goose
- mandrill
- maned sloth
- mountain zebra
- polar bear
- red panda
- sloth bear
- takin
- yak
- great white shark
- American crocodile
- dingo
- king cobra

Near-threatened (NT)

May be considered threatened in the near future.

- Examples: American bison
- Asian golden cat
- blue-billed duck
- emperor goose
- emperor penguin
- Eurasian curlew
- jaguar
- leopard
- Larch Mountain salamander
- Magellanic penguin
- maned wolf
- narwhal
- margay
- montane solitary eagle
- Pampas cat
- Pallas's cat
- reddish egret
- white rhinoceros
- striped hyena
- tiger shark
- white eared pheasant

Least concern (LC)

No immediate threat to species' survival.

- Examples: American alligator
- American crow
- Indian peafowl
- olive baboon
- bald eagle
- brown bear
- brown rat

- brown-throated sloth

- Canada goose

- cane toad

- common wood pigeon

- cougar

- common frog

- giraffe

- grey wolf

- house mouse

- wolverine

- human

- palm cockatoo

- mallard

- meerkat

- mute swan

- platypus

- red-billed quelea

- red-tailed hawk

- rock pigeon

- scarlet macaw

- southern elephant seal

- milk shark

- red howler monkey

Criteria for 'Endangered (EN)'

A) Reduction in population size based on any of the following:

1. An observed, estimated, inferred or suspected population size reduction of ≥ 70% over the last 10 years or three generations, whichever is the longer, where the causes of the reduction are clearly reversible AND understood AND ceased, based on (and specifying) any of the following:

    1. direct observation

    2. an index of abundance appropriate for the taxon

3. a decline in area of occupancy, extent of occurrence and/or quality of habitat

4. actual or potential levels of exploitation

5. the effects of introduced taxa, hybridisation, pathogens, pollutants, competitors or parasites.

2. An observed, estimated, inferred or suspected population size reduction of ≥ 50% over the last 10 years or three generations, whichever is the longer, where the reduction or its causes may not have ceased OR may not be understood OR may not be reversible, based on (and specifying) any of (a) to (e) under A1.

3. A population size reduction of ≥ 50%, projected or suspected to be met within the next 10 years or three generations, whichever is the longer (up to a maximum of 100 years), based on (and specifying) any of (b) to (e) under A1.

4. An observed, estimated, inferred, projected or suspected population size reduction of ≥ 50% over any 10 year or three generation period, whichever is longer (up to a maximum of 100 years in the future), where the time period must include both the past and the future, and where the reduction or its causes may not have ceased OR may not be understood OR may not be reversible, based on (and specifying) any of (a) to (e) under A1.

B) Geographic range in the form of either B1 (extent of occurrence) OR B2 (area of occupancy) OR both:

1. Extent of occurrence estimated to be less than 5,000 km², and estimates indicating at least two of a-c:

    1. Severely fragmented or known to exist at no more than five locations.

    2. Continuing decline, inferred, observed or projected, in any of the following:

        1. extent of occurrence

        2. area of occupancy

        3. area, extent and/or quality of habitat

        4. number of locations or subpopulations

        5. number of mature individuals

    3. Extreme fluctuations in any of the following:

        1. extent of occurrence

        2. area of occupancy

        3. number of locations or subpopulations

        4. number of mature individuals

2. Area of occupancy estimated to be less than 500 km², and estimates indicating at least two of a-c:

1. Severely fragmented or known to exist at no more than five locations.

2. Continuing decline, inferred, observed or projected, in any of the following:

    1. extent of occurrence

    2. area of occupancy

    3. area, extent and/or quality of habitat

    4. number of locations or subpopulations

    5. number of mature individuals

3. Extreme fluctuations in any of the following:

    1. extent of occurrence

    2. area of occupancy

    3. number of locations or subpopulations

    4. number of mature individuals

C) Population estimated to number fewer than 2,500 mature individuals and either:

1. An estimated continuing decline of at least 20% within five years or two generations, whichever is longer, (up to a maximum of 100 years in the future) OR

2. A continuing decline, observed, projected, or inferred, in numbers of mature individuals AND at least one of the follow (a-b):

    1. Population structure in the form of one of the following:

        1. no subpopulation estimated to contain more than 250 mature individuals, OR

        2. at least 95% of mature individuals in one subpopulation

    2. Extreme fluctuations in number of mature individuals

D) Population size estimated to number fewer than 250 mature individuals.

E) Quantitative analysis showing the probability of extinction in the wild is at least 20% within 20 years or five generations, whichever is the longer (up to a maximum of 100 years).

1. Near-critically endangered.

2. Particularly sensitive to poaching levels.

3. Near-endangered due to poaching.

4. May vary according to levels of tourism.

5. Varies according to female populations.

United States

There is data from the United States that shows a correlation between human populations and threatened and endangered species. Using species data from the Database on the Economics and Management of Endangered Species (DEMES) database and the period that the Endangered Species Act (ESA) has been in existence, 1970 to 1997, a table was created that suggests a positive relationship between human activity and species endangerment. As early as the 1800s, humans began noticing the decline of certain species of animals in their usual habitats. An example is the whooping crane. Once abundant from Canada to Mexico, it was estimated in 1941 that only 16 birds remained in the wild. Another early example of mankind noticing the extinction of species was the introduction of "kudzu" in the southern United States. This fast-growing plant took over the south, growing on plants and trees and squelching the life out of them.

Endangered Species Act

Under the Endangered Species Act in the United States, species may be listed as "endangered" or "threatened". The Salt Creek tiger beetle (*Cicindela nevadica lincolniana*) is an example of an endangered subspecies protected under the ESA. The US Fish and Wildlife Service as well as the National Marine Fisheries Service are held responsible for classifying and protecting endangered species, and adding a particular species to the list can be a long, controversial process (Wilcove & Master, 2008, p. 414).

"Endangered" in relation to "threatened" under the ESA.

Some endangered species laws are controversial. Typical areas of controversy include: criteria for placing a species on the endangered species list and criteria for removing a species from the list once its population has recovered; whether restrictions on land development constitute a "taking" of land by the government; the related question of whether private landowners should be compensated for the loss of uses of their lands; and obtaining reasonable exceptions to protection laws. Also lobbying from hunters and various industries like the petroleum industry, construction industry, and logging, has been an obstacle in establishing endangered species laws.

The Bush administration lifted a policy that required federal officials to consult a wildlife expert before taking actions that could damage endangered species. Under the Obama administration, this policy has been reinstated.

Being listed as an endangered species can have negative effect since it could make a species more desirable for collectors and poachers. This effect is potentially reducible, such as in China where commercially farmed turtles may be reducing some of the pressure to poach endangered species.

Another problem with the listing species is its effect of inciting the use of the "shoot, shovel, and shut-up" method of clearing endangered species from an area of land. Some landowners currently may perceive a diminution in value for their land after finding an endangered animal on it. They have allegedly opted to silently kill and bury the animals or destroy habitat, thus removing the problem from their land, but at the same time further reducing the population of an endangered species. The effectiveness of the Endangered Species Act – which coined the term «endangered species» – has been questioned by business advocacy groups and their publications but is nevertheless widely recognized by wildlife scientists who work with the species as an effective recovery tool. Nineteen species have been delisted and recovered and 93% of listed species in the northeastern United States have a recovering or stable population.

Currently, 1,556 known species in the world have been identified as near extinction or endangered and are under protection by government law. This approximation, however, does not take into consideration the number of species threatened with endangerment that are not included under the protection of such laws as the Endangered Species Act. According to NatureServe's global conservation status, approximately thirteen percent of vertebrates (excluding marine fish), seventeen percent of vascular plants, and six to eighteen percent of fungi are considered imperiled. Thus, in total, between seven and eighteen percent of the United States' known animals, fungi and plants are near extinction. This total is substantially more than the number of species protected in the United States under the Endangered Species Act.

Bald eagle

American bison

Ever since mankind began hunting to preserve itself, over-hunting and fishing has been a large and dangerous problem. Of all the species who became extinct due to interference from mankind, the dodo, passenger pigeon, great auk, Tasmanian tiger and Steller's sea cow are some of the more well known examples; with the bald eagle, grizzly bear, American bison, Eastern timber wolf and sea turtle having been hunted to near-extinction. Many began as food sources seen as necessary for survival but became the target of sport. However, due to major efforts to prevent extinction, the bald eagle, or *Haliaeetus leucocephalus* is now under the category of Least Concern on the red list. A present-day example of the over-hunting of a species can be seen in the oceans as populations of certain whales have been greatly reduced. Large whales like the blue whale, bowhead whale, finback whale, gray whale, sperm whale and humpback whale are some of the eight whales which are currently still included on the Endangered Species List. Actions have been taken to attempt reduction in whaling and increase population sizes, including prohibiting all whaling in United States waters, the formation of the CITES treaty which protects all whales, along with the formation of the International Whaling Commission (IWC). But even though all of these movements have been put in place, countries such as Japan continue to hunt and harvest whales under the claim of "scientific purposes". Over-hunting, climatic change and habitat loss leads in landing species in endangered species list and could mean that extinction rates could increase to a large extent in the future.

Invasive Species

The introduction of non-indigenous species to an area can disrupt the ecosystem to such an extent that native species become endangered. Such introductions may be termed alien or invasive species. In some cases the invasive species compete with the native species for food or prey on the natives. In other cases a stable ecological balance may be upset by predation or other causes leading to unexpected species decline. New species may also carry diseases to which the native species have no resistance.

Conservation

The dhole, Asia's most endangered top predator, is on the edge of extinction.

Captive Breeding

Captive breeding is the process of breeding rare or endangered species in human controlled environments with restricted settings, such as wildlife preserves, zoos and other conservation facilities. Captive breeding is meant to save species from extinction and so stabilize the population of the species that it will not disappear.

This technique has worked for many species for some time, with probably the oldest known such instances of captive mating being attributed to menageries of European and Asian rulers, an example being the Père David's deer. However, captive breeding techniques are usually difficult to implement for such highly mobile species as some migratory birds (e.g. cranes) and fishes (e.g. hilsa). Additionally, if the captive breeding population is too small, then inbreeding may occur due to a reduced gene pool and reduce immunity.

In 1981, the Association of Zoos and Aquariums (AZA) created a Species Survival Plan (SSP) in order to help preserve specific endangered and threatened species through captive breeding. With over 450 SSP Plans, there are a number of endangered species that are covered by the AZA with plans to cover population management goals and recommendations for breeding for a diverse and healthy population, created by Taxon Advisory Groups. These programs are commonly created as a last resort effort. SSP Programs regularly participate in species recovery, veterinary care for wildlife disease outbreaks, and a number of other wildlife conservation efforts. The AZA's Species Survival Plan also has breeding and transfer programs, both within and outside of AZA - certified zoos and aquariums. Some animals that are part of SSP programs are giant pandas, lowland gorillas, and California condors.

Private Farming

Black rhino

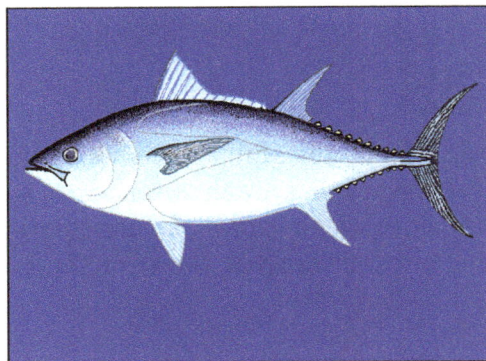

Southern bluefin tuna

Whereas poaching substantially reduces endangered animal populations, legal, for-profit, private farming does the opposite. It has substantially increased the populations of the southern black rhinoceros and southern white rhinoceros. Dr Richard Emslie, a scientific officer at the IUCN, said of such programs, "Effective law enforcement has become much easier now that the animals are largely privately owned... We have been able to bring local communities into the conservation programmes. There are increasingly strong economic incentives attached to looking after rhinos rather than simply poaching: from Eco-tourism or selling them on for a profit. So many owners are keeping them secure. The private sector has been key to helping our work."

Conservation experts view the effect of China's turtle farming on the wild turtle populations of China and South-Eastern Asia – many of which are endangered – as "poorly understood". Although they commend the gradual replacement of turtles caught wild with farm-raised turtles in the marketplace – the percentage of farm-raised individuals in the "visible" trade grew from around 30% in 2000 to around 70% in 2007 – they worry that many wild animals are caught to provide farmers with breeding stock. The conservation expert Peter Paul van Dijk noted that turtle farmers often believe that animals caught wild are superior breeding stock. Turtle farmers may, therefore, seek and catch the last remaining wild specimens of some endangered turtle species.

In 2009, researchers in Australia managed to coax southern bluefin tuna to breed in landlocked tanks, raising the possibility that fish farming may be able to save the species from overfishing.

Countries with Endangered Animals

Around the world hundreds of thousands of species are lost to extinction, many of them only discovered as remains, after they are gone. Thus, not only biological variability, but also genetic diversity, and perhaps sources of livelihood for future generations are lost. An endangered species is a species that may become extinct in the near future. Throughout history millions of species have disappeared, due to natural processes. In the past 300 years, however, humans have increased the rate of extinction.

For some plant and animal species, living seems to be a daily hazard. And humans seem to pose the biggest threat. Ecological disasters, hunting/poaching, deforestation and other consequences of human action causes damage to the food chain, breeding grounds, and habitat.

Threatened Species

Threatened species are any species (including animals, plants, fungi, etc.) which are vulnerable to endangerment in the near future. Species that are threatened are sometimes characterised by the population dynamics measure of *critical depensation*, a mathematical measure of biomass related to population growth rate. This quantitative metric is one method of evaluating the degree of endangerment.

IUCN Definition

The International Union for Conservation of Nature (IUCN) is the foremost authority on threatened species, and treats threatened species not as a single category, but as a group of three categories, depending on the degree to which they are threatened:

- Vulnerable species

- Endangered species

- Critically endangered species

Less-than-threatened categories are near threatened, least concern, and the no longer assigned category of conservation dependent. Species which have not been evaluated (NE), or do not have sufficient data (data deficient) also are not considered "threatened" by the IUCN.

The three categories of the threatened species IUCN Red List.

Although *threatened* and *vulnerable* may be used interchangeably when discussing IUCN categories, the term *threatened* is generally used to refer to the three categories (critically endangered, endangered and vulnerable), while *vulnerable* is used to refer to the least at risk of those three categories. They may be used interchangeably in most contexts however, as all vulnerable species are threatened species (*vulnerable* is a category of *threatened species*); and, as the more at-risk categories of threatened species (namely *endangered* and *critically endangered*) must, by definition, also qualify as vulnerable species, all threatened species may also be considered vulnerable.

Threatened species are also referred to as a red-listed species, as they are listed in the IUCN Red List of Threatened Species.

Subspecies, populations and stocks may also be classified as threatened.

United States Definition

Under the Endangered Species Act in the United States, "threatened" is defined as "any species which is likely to become an endangered species within the foreseeable future throughout all or a significant portion of its range". It is the less protected of the two protected categories. The Bay checkerspot butterfly (*Euphydryas editha bayensis*) is an example of a threatened subspecies protected under the ESA.

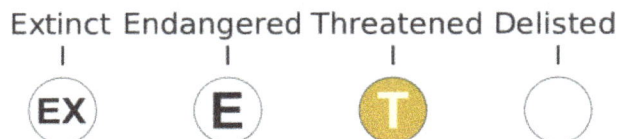

"Threatened" in relation to "endangered" under the ESA.

Within the U.S., state wildlife agencies have the authority under the ESA to manage species which are considered endangered or threatened within their state but not within all states, and which therefore are not included on the national list of endangered and threatened species. For example, the trumpeter swan (*Cygnus buccinator*) is threatened in the state of Minnesota, while large populations still remain in Canada and Alaska.

Australian Definition

The Commonwealth of Australia has legislation for categorising and protecting endangered species, namely the Environment Protection and Biodiversity Conservation Act 1999, which is known in short as the EPBC Act. This act has six categories; extinct, extinct in the wild, critically endangered, endangered, vulnerable, and conservation dependent, defined in Section 179 of the act, and could be summarised as;

- "Extinct" - "no reasonable doubt that the last member of the species has died",

- "Extinct in the wild" - "known only to survive in cultivation" and "despite exhaustive surveys" has not been seen in the wild,

- "Critically endangered" - "extremely high risk of extinction in the wild in the immediate future",

- "Endangered" - "very high risk of extinction in the wild in the near future",

- "Vulnerable" - "high risk of extinction in the wild in medium-term future", and

- "Conservation dependent" - "focus of a specific conservation program" without which the species would enter one of the above categories.

The EPBC Act also recognises and protects threatened ecosystems such as plant communities, and Ramsar Convention wetlands used by migratory birds.

Individual states and territories of Australia are bound under the EPBC Act, but may also have legislation which gives further protection to certain species, for example Western Australia's Wildlife Conservation Act 1950. Some species, such as Lewin's rail (*Lewinia pectoralis*), are not listed as threatened species under the EPBC Act, but they may be recognised as threatened by individual states or territories.

Keystone Species

A keystone species is a species that has a disproportionately large effect on its environment relative to its abundance. Such species are described as playing a critical role in maintaining the structure of an ecological community, affecting many other organisms in an ecosystem and helping to determine the types and numbers of various other species in the community.

The role that a keystone species plays in its ecosystem is analogous to the role of a keystone in an arch. While the keystone is under the least pressure of any of the stones in an arch, the arch still collapses without it. Similarly, an ecosystem may experience a dramatic shift if a keystone species is removed, even though that species was a small part of the ecosystem by measures of biomass or productivity. It became a popular concept in conservation biology. Although the concept is valued as a descriptor for particularly strong inter-species interactions, and it has allowed easier communication between ecologists and conservation policy-makers, it has been criticized for oversimplifying complex ecological systems.

The jaguar, an example of a keystone species

History

Cluster of ochre sea stars (*Pisaster ochraceus*) - keystone predator.

Aggregation of California mussels (*Mytilus californianus*) - prey species.

The concept of the keystone species was introduced in 1969 by Robert T. Paine, a professor of zoology at the University of Washington. Paine developed the concept to explain his observations and experiments on the relationship between intertidal invertebrates. In his 1966 paper, *Food Web Complexity and Species Diversity*, Paine described such a system in Makah Bay in Washington. In his follow-up 1969 paper, Paine proposed the keystone species concept, using *Pisaster ochraceus*, a species of starfish, and *Mytilus californianus*, a species of mussel, as a primary example. The concept became popular in conservation, and was deployed in a range of contexts and mobilized to engender support for conservation.

Examples

Given that there are many historical definitions of the keystone species concept, and without a consensus on its exact definition, a list of examples best illustrates the concept of keystone species.

A classic keystone species is a small predator that prevents a particular herbivorous species from eliminating dominant plant species. Since the prey numbers are low, the keystone predator numbers can be even lower and still be effective. Yet without the predators, the herbivorous prey would explode in numbers, wipe out the dominant plants, and dramatically alter the character of the ecosystem. The exact scenario changes in each example, but the central idea remains that through a chain of interactions, a non-abundant species has an out-sized impact on ecosystem functions. One example is the herbivorous weevil *Euhrychiopsis lecontei* and its suggested keystone effects on aquatic plant species diversity by foraging on nuisance Eurasian watermilfoil.

Similarly, the wasp species *Agelaia vicina* has been labeled a keystone species due to their unparalleled nest size, colony size, and high rate of brood production. The diversity of their prey and the quantity necessary to sustain their high rate of growth have a direct impact on local neighboring species.

Predators

As was described by Dr. Robert Paine in his classic 1966 paper, some sea stars (e.g., *Pisaster ochraceus*) may prey on sea urchins, mussels, and other shellfish that have no other natural predators. If the sea star is removed from the ecosystem, the mussel population explodes uncontrollably, driving out most other species, while the urchin population annihilates coral reefs.

Sea urchins like this purple sea urchin can damage kelp forests by chewing through kelp holdfasts

The sea otter is an important predator of sea urchins.

Similarly, sea otters protect kelp forests from damage by consuming sea urchins. Kelp "roots", called holdfasts, are merely anchors, and do not perform similar roles to the roots of terrestrial plants, which form large networks that acquire nutrients from the soil. In the absence of sea otters, sea urchins are released from predation pressure, increasing in abundance. Sea urchins rapidly consume nearshore kelp, severing the structures at the base. Where sea otters are present, sea urchins tend to be small and limited to crevices. Large nearshore kelp forests proliferate and serve as important habitat for a number of other species. Kelp also increase the productivity of the near-shore ecosystem through the addition of large quantities of secondary production.

These creatures need not be apex predators. Sea stars are prey for sharks, rays, and sea anemones. Sea otters are prey for orca.

The jaguar, whose numbers in Central and South America have been classified as near threatened, acts as a keystone predator by its widely varied diet, helping to balance the mammalian jungle ecosystem with its consumption of 87 different species of prey. The gray wolf is another known keystone species, as is the lion.

Mutualists

Keystone mutualists are organisms that participate in mutually beneficial interactions, the loss of which would have a profound impact upon the ecosystem as a whole. For example, in the Avon Wheatbelt region of Western Australia, there is a period of each year when *Banksia prionotes* (acorn banksia) is the sole source of nectar for honeyeaters, which play an important role in pollination of numerous plant species. Therefore, the loss of this one species of tree would probably cause the honeyeater population to collapse, with profound implications for the entire ecosystem. Another example is frugivores such as the cassowary, which spreads the seeds of many different trees, and some will not grow unless they have been through a cassowary.

Engineers

Beaver dam, an animal construction which has a transformative effect on the environment.

Although the terms 'keystone' and 'engineer' are used interchangeably, the latter is better understood as a subset of keystone species. In North America, the prairie dog is an ecosystem engineer. Prairie dog burrows provide the nesting areas for mountain plovers and burrowing owls. Prairie dog tunnel systems also help channel rainwater into the water table to prevent runoff and erosion, and can also serve to change the composition of the soil in a region by increasing aeration and reversing soil compaction that can be a result of cattle grazing. Prairie dogs also trim the vegetation around their colonies, perhaps to remove any cover for predators. Even grazing species such as plains bison, pronghorn, and mule deer have shown a proclivity for grazing on the same land used by prairie dogs. It is believed that they prefer the plant community which results after prairie dogs have foraged through the area.

Another well known ecosystem engineer, or keystone species, is the beaver, which transforms its territory from a stream to a pond or swamp. Beavers affect the environment first altering the edges of riparian areas by cutting down older trees to use for their dams. This allows younger trees to take their place. Beaver dams alter the riparian area they are established in. Depending on topography, soils, and many factors, these dams change the riparian edges of streams and rivers into wetlands, meadows, or riverine forests. These dams have shown to be beneficial to myriad species including amphibians, salmon, and song birds.

In the African savanna, the larger herbivores, especially the elephants, shape their environment. The elephants destroy trees, making room for the grass species. Without these animals, much of the savannah would turn into woodland.

On the Great Barrier Reef, Australia studies have found the parrotfish on the Great Barrier Reef is the sole species, within thousands of species of reef fish that consistently scrapes and cleans the coral on the reef. Without these animals, the Great Barrier Reef would be under severe strain.

Indicator Species

An indicator species is any biological species that defines a trait or characteristic of the environment. For an example, a species may delineate an ecoregion or indicate an environmental condition such as a disease outbreak, pollution, species competition or climate change. Indicator species can be among the most sensitive species in a region, and sometimes act as an early warning to monitoring biologists.

Animal species have been used for indicators for decades to collect information about the many regions. Vertebrate are used as population trends and habitat for other species. Species identification is very important for the conservation of biodiversity. Approximately 1.9 million species have been identified, but there are 3 to 100 million species. Some of them haven't been studied. There are new species every year that are unknown and are still being discovered each year. Indicator species serve as measured environmental conditions.

Indicator species are also known as sentinel organisms, i.e. organisms which are ideal for biomonitoring. Organisms such as oysters, clams, and cockles have been extensively used as biomonitors in marine and estuarine environments. For example, the Mussel Watch Programme is a world-wide project using mussels to assess environmental impacts on coastal waters. Their well-documented feeding habits, stationary condition and their role as integral parts of the food chain are some of the main reasons why oysters and mussels are widely used biomonitors. A considerable amount of contaminant concentrations are found in the surficial sediments (i.e. the finer-grained particulate matter, usually muds, silts or clays) of marine and estuarine environments. A major physical process governing the transport of fine particulate material and associated particle-bound contaminants in estuarine environments is resuspe nsion. Strong winds create surface waves, which, in shallow water (<5m), project energy to the water-sediment interface resulting in resuspension of fine sediment from the upper layers of the estuary floor. Once in suspension, fine material may be transported by tidal currents to other parts of the estuary and possibly to the ocean during multiple reworking phases. Mussels and oysters are filter feeders and therefore uptake is by ingestion of particulates in the water column. Sediment resuspension is thus very important in the bioaccumulation process which aids the evaluation of possible adverse biological effects of sedimentary contaminants in marine and estuarine environments.

Examples

- Stoneflies: indicate high oxygen water

  Stoneflies spend the majority of their lives as nymphs. Many species require a high concentration of dissolved oxygen and are found in clean swift streams with gravel or stone bottom.

- Mosses: indicate acidic soil

- Greasewood: indicates saline soil

  Greasewood grows on dry, sunny flat valley bottoms on ephemeral stream channels. It is one of the dominant plants throughout Great Basin and Mojave Desert. In high saline areas, greasewood grows in nearly pure stands.

- Lichens: some species indicate low air pollution

  Lichens as a group have a worldwide distribution and grow almost on any surface, for example soil, bark, roof tiles or stone. Because lichens get all their nutrients from the air, many species are very sensitive to air pollution.

- Fungi: high conservation value, old-growth forests

  Wood-decay fungi are the basis of saprotrophic species communities in forest ecosystems, and sensitive to intensive forest management. Their diversity correlates well with insect diversity, and indicates continuum of dead wood at the stand or landscape level. Fungi, bracket fungi in particular, are used as indicators of conservation value in forest inventories, particularly in Nordic countries and boreal Russia.

- Mollusca: numerous bivalve molluscs indicate water pollution status

  Mollusca, and quite often bivalve molluscs are used as bioindicators to monitor the health of an aquatic environment, either fresh- or seawater. Their population status or structure, physiology, behaviour or their content of certain elements or compounds can reveal the contamination status of any aquatic ecosystem. They are extremely useful as they are sessile - which means they are closely representative of the environment where they are sampled or placed (caging) -, and they are breathing water all along the day, exposing their gills and internal tissues: bioaccumulation. One of the most famous project in that field is the Mussel Watch Programme but today they are used worldwide for that purpose (Ecotoxicology).

- Tubifex worms: indicate nonpotable, stagnant, oxygen-poor water
- Agave lechuguilla: an important indicator species in the Chihuahuan Desert

Definitions

Lindenmayer *et al.* suggest seven alternative definitions of indicator species:

1. a species whose presence indicates the presence of a set of other species and whose absence indicates the lack of that entire set of species;

2. a keystone species, which is a species whose addition to or loss from an ecosystem leads to major changes in abundance or occurrence of at least one other species

3. a species whose presence indicates human-created abiotic conditions such as air or water pollution (often called a pollution indicator species)

4. a dominant species that provides much of the biomass or number of individuals in an area

5. a species that indicates particular environmental conditions such as certain soil or rock types

6. a species thought to be sensitive to and therefore to serve as an early warning indicator of environmental changes such as global warming or modified fire regimes (sometimes called a bioindicator species)

7. a management indicator species, which is a species that reflects the effects of a disturbance regime or the efficacy of efforts to mitigate disturbance effects.

Type 1, 2, and 4 have been proposed as indicators of biological diversity and types 3, 5, 6, and 7 as indicators of abiotic conditions and/or changes in ecological processes.

Umbrella Species

Umbrella species are species selected for making conservation-related decisions, typically because protecting these species indirectly protects the many other species that make up the ecological community of its habitat. Species conservation can be subjective because it is hard to determine the status of many species. With millions of species of concern, the identification of selected *keystone species, flagship species* or *umbrella species* makes conservation decisions easier. Umbrella species can be used to help select the locations of potential reserves, find the minimum size of these conservation areas or reserves, and to determine the composition, structure and processes of ecosystems.

Definitions

Two commonly used definitions:

- A: "A wide-ranging species whose requirements include those of many other species"

- B: A species with large area requirements for which protection of the species offers protection to other species that share the same habitat

Other descriptions include:

- A: "The protection of umbrella species automatically extends protection to other species. i.e. spotted owl and old growth trees"

- B: "Traditional umbrella species, relatively large-bodied and wide-ranging species of higher vertebrates"

Use in Landuse Management

The use of umbrella species as a conservation tool is highly debated. The term was first used by Wilcox (1984) who defined an umbrella species as one whose minimum area requirements are at least as comprehensive of the rest of the community for which protection is sought though the establishment and management of a protected area.

Some scientists have found that the umbrella effect provides a simpler way to manage ecological communities. Others feel that a combination of other tools establish better land management reserves to help protect more species than just using umbrella species alone. Individual invertebrate species can be good umbrella species because they can protect older, unique ecosystems. There have been cases where umbrella species have protected a large amount of area which has been beneficial to surrounding species such as the northern spotted owl.

Currently research is being done on land management decisions based on using umbrella species to protect habitat of specific species as well as other organisms in the area. Dunk, Zielinski and Welsh (2006) reported that the reserves in Northern California (Klamath-Siskiyou forests), set aside for the northern spotted owl, also protect mollusks and salamanders within that habitat. According to their conclusions, the reserves set aside for the northern spotted owl "serve as a reasonable coarse-filter umbrella species for the taxa [they] evaluated," which were the mollusks and salamanders.

Use in the Endangered Species Act (USA)

The Bay checkerspot butterfly has been on the Endangered Species List since 1987 and is still currently listed. Launer and Murphy (1994) tried to determine whether this butterfly could be considered an umbrella species in protecting the native grassland it inhabits. They discovered that the Endangered Species Act (ESA) has a loophole to eliminate federally protected plants that reside on private property. However, the California Environmental Quality Act (CEQA) reinforces state conservation regulations. Using the ESA to protect termed umbrella species and their habitats can be controversial because they are not as reinforced in some states as others (such as California) to protect overall biodiversity.

Examples of Umbrella Species

1. Northern spotted owls and old growth forest : ex. Molluscs and salamanders are within the protective boundaries of the northern spotted owl.

2. Bay checkerspot butterfly and grasslands

3. Amur Tigers in the Russian Far East are considered Umbrella/Keystone Species due to their impact on the deer and boar in their ecosystem

Flagship Species

The concept of flagship species has its genesis in the field of conservation biology. The flagship species concept holds that by raising the profile of a particular species, it can successfully leverage more support for biodiversity conservation at large in a particular context.

Project logo showing the use of the Zanzibar red colobus as the flagship species for a conservation organization in Zanzibar, Tanzania.

Logo showing the use of the Eurasian lynx as the flagship species for a protected area in Poland

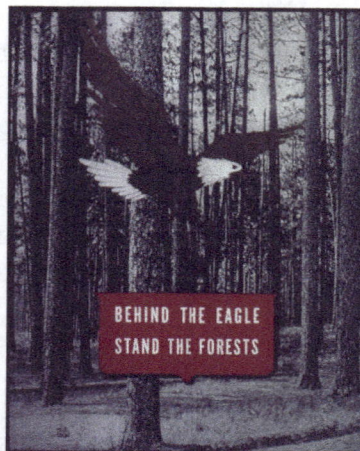

Poster logo showing the use of the bald eagle as the flagship species for forests in the United States

Display showing the use of the tiger as the flagship species for a campaign at Berijam lake in Kodaikanal, India

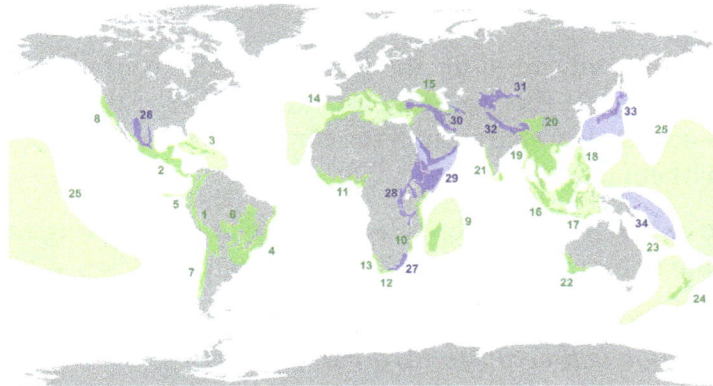

The twenty-five biodiversity hotspots (green) as indicated in Myers, N., et al. (2000) "Biodiversity hotspots for conservation priorities." Nature 403:853–858. Hotspots are a conservation flagship scheme used to raise awareness and funds for the regions of the world with more endemic species and which are under larger threat of disappearing

500 Tanzanian shillings bank note showing the use of the African buffalo as a flagship species for the country's wildlife.

5000 Tanzanian shillings bank note showing the use of the Black rhinoceros as a flagship species for the country's wildlife.

10000 Tanzanian shillings bank note showing the use of the African savanna elephant as a flagship species for the country's wildlife.

Definitions

Several definitions have been advanced for the flagship species concept and for some time been there has been confusion even in the academic literature. Most of the latest definitions focus on the strategic and socio-economical character of the concept, with a recent publication establishing a clear link with the marketing field.

- "a species used as the focus of a broader conservation marketing campaign based on its possession of one or more traits that appeal to the target audience."

- *species that have the ability to capture the imagination of the public and induce people to support conservation action and/or to donate funds*

- *popular, charismatic species that serve as symbols and rallying points to stimulate conservation awareness and action*

The term flagship is linked to the metaphor of representation. In its popular usage, flagships are viewed as ambassadors or icons for a conservation project or movement.

However, more recently, work in the field of microbiology has started to use the concept of flagship species in a distinct way. This work relates to the biogeography of micro-organisms and uses particular species because "eyecatching "flagships" with conspicuous size and/or morphology are the best distribution indicators".

Examples

Examples of flagship species include the Bengal tiger (*Panthera tigris*), the giant panda (*Ailuropoda melanoleuca*), the Golden lion tamarin (*Leontopithecus rosalia*), the African elephant (*Loxodonta sp.*) and Asian elephant (*Elephas maximus*).

Flagship species can represent an environmental feature (e.g. a species or ecosystem), cause (e.g. climate change or ocean acidification), organization (e.g. NGO or government department) or geographic region (e.g. state or protected area).

History

The flagship species concept appears to have become popular around the mid 1980s within the debate on how to prioritise species for conservation. The first widely available references to use the flagship concept applied it to both neotropical primates and African elephants and rhinos, in the typical mammal centric approach that still dominates how the concept is used today

The use of the concept has been largely dominated by large bodied species, especially mammals, although species from other taxonomic groups have occasionally been used

Selection

Flagship species can be selected according to many different characteristics depending on what is valued by the audience they try to target. This is best illustrated by the differences in recommendations made for flagship species selection targeting different target audiences such as local communities. and tourists.

Limitations

Several limitations have been recognized concerning the use of flagship species:

- The use of flagship species can skew the management and conservation priorities in their favour and to the detriment of more threatened species

- The managements of different flagships can conflict

- The disappearance of the flagship can have negative impacts on the attitudes of the conservation stakeholders

Flagships and Conflict

A major challenge for the deployment of several flagship species in non-Western contexts is that they may come into conflict with local communities, thereby jeopardizing well-intended conservation actions. This has been termed 'flagship mutiny' and is exemplified by the Asian elephant in countries where there is human-elephant conflict.

Other Types of Conservation Flagships

Conservation flagships can also appear at broader levels, for example as ecosystems (such as coral reefs or rainforests or protected areas (Serengeti or Yellowstone). A number of recent initiatives has developed new conservation flagships based on conservation values of particular areas or species. Examples of these are the EDGE project run by the Zoological Society of London and the Hotspots run by Conservation International.

Introduced Species

An introduced, alien, exotic, non-indigenous, or non-native species, or simply an introduction, is a species living outside its native distributional range, which has arrived there by human activity, either deliberate or accidental. Non-native species can have various effects on the local ecosystem. Introduced species that become established and spread beyond the place of introduction are called invasive species. Some have a negative effect on a local ecosystem. Some introduced species may have no negative effect or only minor impact. Some species have been introduced intentionally to combat pests. They are called biocontrols and may be regarded as beneficial as an alternative to pesticides in agriculture for example. In some instances the potential for being beneficial or detrimental in the long run remains unknown.

Sweet clover (*Melilotus sp.*), introduced and naturalized to the U.S. from Eurasia as a forage and cover crop.

The effects of introduced species on natural environments have gained much scrutiny from scientists, governments, farmers and others.

The terminology associated with introduced species is now in flux for various reasons. Other terms with somewhat similar meanings) with *introduced* are *acclimatized*, *adventive*, *naturalized*, and *immigrant* but those terms refer to a subset of introduced species: those that have become established and can reproduce without human assistance. The term invasive refers only to those species that become established and spread beyond the place of introduction. For practical purposes, this term is applied only to invasive species that cause damage.

In the broadest and most widely used sense, an introduced species is synonymous with *non-native* and therefore applies as well to most garden and farm organisms; these adequately fit the basic definition given above. However, some sources add to that basic definition "and are now reproducing in the wild," which removes from consideration as *introduced* all of those species raised or grown in gardens or farms that do not survive without tending by people. With respect to plants, these latter are in this case defined as either *ornamental* or *cultivated* plants.

The following definition from the United States Environmental Protection Agency (EPA), although perhaps lacking ecological sophistication, is more typical: *introduced species* "A species that has been intentionally or inadvertently brought into a region or area. Also called an exotic or non-native species." Introduction of a species outside its native range is all that is required to be qualified as an "introduced species" such that one can distinguish between introduced species that may not occur except in cultivation, under domestication or captivity whereas others become established outside their native range and reproduce without human assistance. Such species might be termed "naturalized", "established", "wild non-native species". If they further spread beyond the place of introduction they are called "invasive". The transition from introduction, to establishment and to invasion has been described in the context of plants. Introduced species are essentially "non-native" species. Invasive species are those introduced species that spreadwidely or quickly and cause harm, be that to the environment, human health, other valued resources or the economy. There have been calls from scientists to consider a species "invasive" only in terms of their spread and reproduction rather than the harm they may cause.

According to a practical definition, an invasive species is one that has been introduced and become a pest in its new location, spreading (invading) by natural means. The term is used to imply both a sense of urgency and actual or potential harm. For example, U.S. Executive Order 13112 (1999) defines "invasive species" as "an alien species whose introduction does or is likely to cause economic or environmental harm or harm to human health". The biological definition of invasive species, on the other hand, makes no reference to the harm they may cause, only to the fact that they spread beyond the area of original introduction.

Although some argue that "invasive" is a loaded word and harm is difficult to define, the fact of the matter is that organisms have and continue to be introduced to areas in which they are not native, sometimes with but usually without much regard to the harm that could result.

From a regulatory perspective, it is neither desirable nor practical to list as undesirable or outright ban all non-native species (although the State of Hawaii has adopted an approach that comes close to this). Regulations require a definitional distinction between non-natives that are deemed especially onerous and all others. Introduced *pest species* that are *officially listed* as invasive, best fit the definition of an *invasive species*. Early detection and rapid response is the most effective strategy for regulating a pest species and reducing economic and environmental impacts of an introduction

In Great Britain, the Wildlife and Countryside Act 1981 prevents the introduction of any animal not naturally occurring in the wild or any of a list of both animals or plants introduced previously and proved to be invasive.

Nature of Introductions

By definition, a species is considered "introduced" when its transport into an area outside of its native range is human mediated. Introductions by humans can be described as either intentional or accidental. Intentional introductions have been motivated by individuals or groups who either (1) believe that the newly introduced species will be in some way beneficial to humans in its new location or, (2) as is the case with pythons in the Everglades, species are introduced intentionally but with no regard to the potential impact. Unintentional or accidental introductions are most often a byproduct of human movements, and are thus unbound to human motivations. Subsequent range expansion of introduced species may or may not involve human activity.

Intentional Introductions

Species that humans intentionally transport to new regions can subsequently become successfully established in two ways. In the first case, organisms are purposely released for establishment in the wild. It is sometimes difficult to predict whether a species will become established upon release, and if not initially successful, humans have made repeated introductions to improve the probability that the species will survive and eventually reproduce in the wild. In these cases it is clear that the introduction is directly facilitated by human desires.

In the second case, species intentionally transported into a new region may escape from captive or cultivated populations and subsequently establish independent breeding populations. Escaped organisms are included in this category because their initial transport to a new region is human motivated.

Perhaps the most common motivation for introducing a species into a new place is that of economic gain. Examples of species introduced for the purposes of benefiting agriculture, aquaculture or other economic activities are widespread. Eurasian carp was first introduced to the United States as a potential food source. The apple snail was released in Southeast Asia with the intent that it be used as a protein source, and subsequently to places like Hawaii to establish a food industry. In Alaska, foxes were introduced to many islands to create new populations for the fur trade. About twenty species of African and European dung beetles have established themselves in Australia after deliberate introduction by the Australian Dung Beetle Project in an effort to reduce the impact of livestock manure. The timber industry promoted the introduction of Monterey pine (*Pinus radiata*) from California to Australia and New Zealand as a commercial timber crop. These examples represent only a small subsample of species that have been moved by humans for economic interests.

Introductions have also been important in supporting recreation activities or otherwise increasing human enjoyment. Numerous fish and game animals have been introduced for the purposes of sport fishing and hunting (Earthworms as invasive species). The introduced amphibian (*Ambystoma tigrinum*) that threatens the endemic California salamander (*Ambystoma californiense*) was introduced to California as a source of bait for fishermen. Pet animals have also been frequently transported into new areas by humans, and their escapes have resulted in several successful introductions, such as those of feral cats and parrots.

Many plants have been introduced with the intent of aesthetically improving public recreation areas or private properties. The introduced Norway maple for example occupies a prominent status in many of Canada's parks. The transport of ornamental plants for landscaping use has and continues to be a source of many introductions. Some of these species have escaped horticultural control and become invasive. Notable examples include water hyacinth, salt cedar, and purple loosestrife.

In other cases, species have been translocated for reasons of "cultural nostalgia," which refers to instances in which humans who have migrated to new regions have intentionally brought with them familiar organisms. Famous examples include the introduction of starlings to North America by Englishman Eugene Schieffelin, a lover of the works of Shakespeare and the chairman of the American Acclimatization Society, who, it is rumoured, wanted to introduce all of the birds mentioned in Shakespeare's plays into the United States. He deliberately released eighty starlings into Central Park in New York City in 1890, and another forty in 1891.

Yet another prominent example is the introduction of the European rabbit to Australia by one Thomas Austin, a British landowner who had the rabbits released on his estate in Victoria because he missed hunting them. A more recent example is the introduction of the common wall lizard to North America by a Cincinnati boy, George Rau, around 1950 after a family vacation to Italy.

Intentional introductions have also been undertaken with the aim of ameliorating environmental problems. A number of fast spreading plants such as garlic mustard and kudzu have been introduced as a means of erosion control. Other species have been introduced as biological control agents to control invasive species and involves the purposeful introduction of a natural enemy of the target species with the intention of reducing its numbers or controlling its spread.

A special case of introduction is the reintroduction of a species that has become locally endangered or extinct, done in the interests of conservation. Examples of successful reintroductions include wolves to Yellowstone National Park in the U.S., and the red kite to parts of England and Scotland. Introductions or translocations of species have also been proposed in the interest of genetic conservation, which advocates the introduction of new individuals into genetically depauperate populations of endangered or threatened species.

The above examples highlight the intent of humans to introduce species as a means of incurring some benefit. While these benefits have in some cases been realized, introductions have also resulted in unforeseen costs, particularly when introduced species take on characteristics of invasive species.

Non-native species can become such a common part of an environment, culture, and even diet that little thought is given to their geographic origin. For example, soybeans, kiwi fruit, wheat and all livestock except the American bison and the turkey are non-native species to North America. Collectively, non-native crops and livestock comprise 98% of US food. These and other benefits from non-natives are so vast that, according to the Congressional Research Service, they probably exceed the costs.

Unintentional Introductions

Unintentional introductions occur when species are transported by human vectors. For example, three species of rat (the black, Norway and Polynesian) have spread to most of the world as hitchhikers on ships. There are also numerous examples of marine organisms being transported in ballast water, one being the zebra mussel. Over 200 species have been introduced to the San Francisco Bay in this manner making it the most heavily invaded estuary in the world. Increasing rates of human travel are providing accelerating opportunities for species to be accidentally transported into areas in which they are not considered native. There is also the accidental release of the Africanized honey bees (AHB), known colloquially as "killer bees" or Africanized bee to Brazil in 1957 and the Asian carps to the United States. The insect commonly known as the brown marmorated stink bug (*Halyomorpha halys*) was introduced accidentally in Pennsylvania. Another form of unintentional introductions is when an intentionally introduced plant carries a parasite or herbivore with it. Some become invasive, for example the oleander aphid, accidentally introduced with the ornamental plant, oleander.

Introduced Plants

Many non-native plants have been introduced into new territories, initially as either ornamental plants or for erosion control, stock feed, or forestry. Whether an exotic will become an invasive species is seldom understood in the beginning, and many non-native ornamentals languish in the trade for years before suddenly naturalizing and becoming invasive.

Peaches, for example, originated in China, and have been carried to much of the populated world. Tomatoes are native to the Andes. Squash (pumpkins), maize (corn), and tobacco are native to the Americas, but were introduced to the Old World. Many introduced species require continued human intervention to survive in the new environment. Others may become feral, but do not seriously compete with natives, but simply increase the biodiversity of the area.

Dandelions are also introduced species to North America.

A very troublesome marine species in southern Europe is the seaweed *Caulerpa taxifolia*. *Caulerpa* was first observed in the Mediterranean Sea in 1984, off the coast of Monaco. By 1997, it had covered some 50 km². It has a strong potential to overgrow natural biotopes, and represents a major risk for sublittoral ecosystems. The origin of the alga in the Mediterranean was thought to be either as a migration through the Suez Canal from the Red Sea, or as an accidental introduction from an aquarium.

Japanese knotweed grows profusely in many nations. Human beings introduced it into many places in the 19th century. It is a source of resveratrol, a dietary supplement.

Introduced Animals

Bear in mind that most introduced species do not become invasive. Examples of introduced animals that have become invasive include the gypsy moth in eastern North America, the zebra mussel and alewife in the Great Lakes, the Canada goose and gray squirrel in Europe, the muskrat in Europe and Asia, the cane toad and red fox in Australia, nutria in North America, Eurasia, and Africa, and the common brushtail possum in New Zealand.

Most Commonly Introduced Species

Some species, such as the brown rat, house sparrow, ring-necked pheasant and European starling, have been introduced very widely. In addition there are some agricultural and pet species that frequently become feral; these include rabbits, dogs, goats, fish, pigs and cats.

Invasive Exotic Diseases

History is rife with the spread of exotic diseases, such as the introduction of smallpox into the indigenous peoples of the Americas by the Spanish, where it obliterated entire populations of indigenous civilizations before they were ever even seen by Europeans.

Problematic exotic disease introductions in the past century or so include the chestnut blight which has almost eliminated the American chestnut tree from its forest habitat, and Dutch elm disease, which has severely reduced the American elm trees in forests and cities.

Diseases may also be vectored by invasive insects such as the Asian citrus psyllid and the bacterial disease citrus greening.

Introduced Species on Islands

Perhaps the best place to study problems associated with introduced species is on islands. Depending upon the isolation (how far an island is located from continental biotas), native island biological communities may be poorly adapted to the threat posed by exotic introductions. Often this can mean that no natural predator of an introduced species is present, and the non-native spreads uncontrollably into open or occupied niche.

An additional problem is that birds native to small islands may have become flightless because of the absence of predators prior to introductions and cannot readily escape danger. The tendency of rails in particular to evolve flightless forms on islands has led to the disproportionate number of extinctions in that family.

The field of island restoration has developed as a field of conservation biology and ecological restoration, a great deal of which deals with the eradication of introduced species.

New Zealand

In New Zealand the largest commercial crop is *Pinus radiata*, the native Californian Monterey pine tree, which grows as well in New Zealand as in California. However, the pine forests are also occupied by deer from North America and Europe and by possums from Australia. All are exotic species and all have thrived in the New Zealand environment. The pines are seen as beneficial while the deer and possums are regarded as serious pests.

Common gorse, originally a hedge plant in Britain, was introduced to New Zealand for the same purpose. Like the Monterey pine, it has shown a favour to its new climate. It is, however, regarded as a noxious plant that threatens to obliterate native plants in much of the country and is hence routinely eradicated, though it can also provide a nursery environment for native plants to reestablish themselves.

Rabbits, introduced as a food source by sailors in the 1800s, have become a severe nuisance to farmers, notably in the South Island. The myxomatosis virus was illegally imported and illegally released, but it had little lasting effect upon the rabbit population other than to make it more resistant to the virus.

Cats, brought by the Māori and later by Europeans, have had a devastating effect upon the native birdlife, particularly as many New Zealand birds are flightless. Feral cats and dogs which were originally brought as pets are also known to kill large numbers of birds. A recent (2006) study in the South Island has shown that even domestic cats with a ready supply of food from their owners may kill hundreds of birds in a year, including natives.

Sparrows, which were brought to control insects upon the introduced grain crops, have displaced native birds as have rainbow lorikeets and cockatoos (both from Australia) which fly free around areas west of Auckland City such as the Waitakere Ranges.

In much of New Zealand, the Australian black swan has effectively eliminated the existence of the previously introduced mute swan.

Two notable varieties of spiders have also been introduced: the white tail spider and the redback spider. Both may have arrived inside shipments of fruit. Until then, the only spider (and the only poisonous animal) dangerous to humans was the native katipo, which is very similar to the redback and interbreed with the more aggressive Australian variety.

Introduced Species on a Planetary Body

It has been hypothesized that invasive species of microbial life could contaminate a planetary body after the former is introduced by a space probe or spacecraft, either deliberately or unintentionally.

Invasive Species

An invasive species is a plant, fungus, or animal species that is not native to a specific location (an introduced species), and which has a tendency to spread to a degree believed to cause damage to the environment, human economy or human health.

Beavers from North America constitute an invasive species in Tierra del Fuego, where they have a substantial impact on landscape and local ecology through their dams.

Kudzu, a Japanese vine species invasive in the southeast United States, growing in Atlanta, Georgia

One study pointed out widely divergent perceptions of the criteria for invasive species among researchers (p. 135) and concerns with the subjectivity of the term "invasive" (p. 136). Some of the alternate usages of the term are below:

- The term as most often used applies to introduced species (also called "non-indigenous" or "non-native") that adversely affect the habitats and bioregions they invade economically, environmentally, or ecologically. Such invasive species may be either plants or animals and may disrupt by dominating a region, wilderness areas, particular habitats, or wildland–urban interface land from loss of natural controls (such as predators or herbivores). This includes non-native invasive plant species labeled as exotic pest plants and invasive exotics growing in native plant communities. It has been used in this sense by government organizations as well as conservation groups such as the International Union for Conservation of Nature (IUCN) and the California Native Plant Society. The European Union defines "Invasive Alien Species" as those that are, firstly, outside their natural distribution area, and secondly, threaten biological diversity. It is also used by land managers, botanists, researchers, horticulturalists, conservationists, and the public for noxious weeds. The kudzu vine (*Pueraria lobata*), Andean Pampas grass (*Cortaderia jubata*), and yellow starthistle (*Centaurea solstitialis*) are examples.

- An alternate usage broadens the term to include indigenous or "native" species along with *non-native* species, that have colonized natural areas (p. 136). Deer are an example, considered to be overpopulating their native zones and adjacent suburban gardens, by some in the Northeastern and Pacific Coast regions of the United States.

- Sometimes the term is used to describe a non-native or introduced species that has become widespread (p. 136). However, not every introduced species has adverse effects on the environment. A nonadverse example is the common goldfish (*Carassius auratus*), which is found throughout the United States, but rarely achieves high densities (p. 136).

Causes

Scientists include species- and ecosystem factors among the mechanisms that when combined, establish invasiveness in a newly introduced species.

Species-based Mechanisms

While all species compete to survive, invasive species appear to have specific traits or specific combinations of traits that allow them to outcompete native species. In some cases, the competition is about rates of growth and reproduction. In other cases, species interact with each other more directly.

Researchers disagree about the usefulness of traits as invasiveness markers. One study found that of a list of invasive and noninvasive species, 86% of the invasive species could be identified from the traits alone. Another study found invasive species tended to have only a small subset of the presumed traits and that many similar traits were found in noninvasive species, requiring other explanations. Common invasive species traits include the following:

- Fast growth

- Rapid reproduction

- High dispersal ability

- Phenotypic plasticity (the ability to alter growth form to suit current conditions)

- Tolerance of a wide range of environmental conditions (Ecological competence)

- Ability to live off of a wide range of food types (generalist)

- Association with humans

- Prior successful invasions

Typically, an introduced species must survive at low population densities before it becomes invasive in a new location. At low population densities, it can be difficult for the introduced species to reproduce and maintain itself in a new location, so a species might reach a location multiple times before it becomes established. Repeated patterns of human movement, such as ships sailing to and from ports or cars driving up and down highways offer repeated opportunities for establishment (also known as a high propagule pressure).

An introduced species might become invasive if it can outcompete native species for resources such as nutrients, light, physical space, water, or food. If these species evolved under great competition or predation, then the new environment may host fewer able competitors, allowing the invader to proliferate quickly. Ecosystems in which are being used to their fullest capacity by native species can be modeled as zero-sum systems in which any gain for the invader is a loss for the native. However, such unilateral competitive superiority (and extinction of native species with increased populations of the invader) is not the rule. Invasive species often coexist with native species for an extended time, and gradually, the superior competitive ability of an invasive species becomes apparent as its population grows larger and denser and it adapts to its new location.

Lantana growing in abandoned citrus plantation; Moshav Sdei Hemed, Israel

An invasive species might be able to use resources that were previously unavailable to native species, such as deep water sources accessed by a long taproot, or an ability to live on previously uninhabited soil types. For example, barbed goatgrass (*Aegilops triuncialis*) was introduced to California on serpentine soils, which have low water-retention, low nutrient levels, a high magnesium/calcium ratio, and possible heavy metal toxicity. Plant populations on these soils tend to show low density, but goatgrass can form dense stands on these soils and crowd out native species that have adapted poorly to serpentine soils.

Invasive species might alters its environment by releasing chemical compounds, modifying abiotic factors, or affecting the behaviour of herbivores, creating a positive or negative impact on other species. Some species, like *Kalanchoe daigremontana*, produce allelopathic compounds, that might have an inhibitory effect on competing species. Other species like *Stapelia gigantea* facilitates the recruitment of seedlings of other species in arid environments by providing appropriate microclimatic conditions and preventing herbivory in early stages of development.

Another examples are *Centaurea solstitialis* (yellow starthistle) and *Centaurea diffusa* (diffuse knapweed). These Eastern European noxious weeds have spread through the western and West Coast states. Experiments show that 8-hydroxyquinoline, a chemical produced at the root of *C. diffusa*, has a negative effect only on plants that have not co-evolved with it. Such co-evolved native plants have also evolved defenses. *C. diffusa* and *C. solstitialis* do not appear in their native habitats to be overwhelmingly successful competitors. Success or lack of success in one habitat does not necessarily imply success in others. Conversely, examining habitats in which a species is less successful can reveal novel weapons to defeat invasiveness.

Changes in fire regimens are another form of facilitation. *Bromus tectorum*, originally from Eurasia, is highly fire-adapted. It not only spreads rapidly after burning but also increases the frequency and intensity (heat) of fires by providing large amounts of dry detritus during the fire season in western North America. In areas where it is widespread, it has altered the local fire regimen so much that native plants cannot survive the frequent fires, allowing *B. tectorum* to further extend and maintain dominance in its introduced range.

Facilitation also occurs where one species physically modifies a habitat in ways that are advantageous to other species. For example, zebra mussels increase habitat complexity on lake floors, providing crevices in which invertebrates live. This increase in complexity, together with the nutrition provided by the waste products of mussel filter-feeding, increases the density and diversity of benthic invertebrate communities.

Ecosystem-based Mechanisms

In ecosystems, the amount of available resources and the extent to which those resources are used by organisms determines the effects of additional species on the ecosystem. In stable ecosystems, equilibrium exists in the use of available resources. These mechanisms describe a situation in which the ecosystem has suffered a disturbance, which changes the fundamental nature of the ecosystem.

When changes such as a forest fire occur, normal succession favors native grasses and forbs. An introduced species that can spread faster than natives can use resources that would have been available to native species, squeezing them out. Nitrogen and phosphorus are often the limiting factors in these situations.

Every species occupies a *niche* in its native ecosystem; some species fill large and varied roles, while others are highly specialized. Some invading species fill niches that are not used by native species, and they also can create new niches. An example of this type can be found within the *Lampropholis delicata* species of skink.

Ecosystem changes can alter species' distributions. For example, edge effects describe what happens when part of an ecosystem is disturbed as when land is cleared for agriculture. The boundary between remaining undisturbed habitat and the newly cleared land itself forms a distinct habitat, creating new winners and losers and possibly hosting species that would not thrive outside the boundary habitat.

One interesting finding in studies of invasive species has shown that introduced populations have great potential for rapid adaptation and this is used to explain how so many introduced species are able to establish and become invasive in new environments. When bottlenecks and founder effects cause a great decrease in the population size, the individuals begin to show additive variance as opposed to epistatic variance. This conversion can actually lead to increased variance in the founding populations which then allows for rapid adaptive evolution. Following invasion events, selection may initially act on the capacity to disperse as well as physiological tolerance to the new stressors in the environment. Adaptation then proceeds to respond to the selective pressures of the new environment. These responses would most likely be due to temperature and climate change, or the presence of native species whether it be predator or prey. Adaptations include changes in morphology, physiology, phenology, and plasticity.

Rapid adaptive evolution in these species leads to offspring that have higher fitness and are better suited for their environment. Intraspecific phenotypic plasticity, pre-adaptation and post-introduction evolution are all major factors in adaptive evolution. Plasticity in populations allows room for changes to better suit the individual in its environment. This is key in adaptive evolution because the main goal is how to best be suited to the ecosystem that the species has been introduced. The ability to accomplish this as quickly as possible will lead to a population with a very high fitness. Pre-adaptations and evolution after the initial introduction also play a role in the success of the introduced species. If the species has adapted to a similar ecosystem or contains traits that happen to be well suited to the area that it is introduced, it is more likely to fare better in the new environment. This, in addition to evolution that takes place after introduction, all determine if the species will be able to become established in the new ecosystem and if it will reproduce and thrive.

Ecology

Traits of Invaded Ecosystems

In 1958, Charles S. Elton claimed that ecosystems with higher species diversity were less subject to invasive species because of fewer available niches. Other ecologists later pointed to highly diverse, but heavily invaded ecosystems and argued that ecosystems with high species diversity were more susceptible to invasion.

This debate hinged on the spatial scale at which invasion studies were performed, and the issue of how diversity affects susceptibility remained unresolved as of 2011. Small-scale studies tended to show a negative relationship between diversity and invasion, while large-scale studies tended to show the reverse. The latter result may be a side-effect of invasives' ability to capitalize on increased resource availability and weaker species interactions that are more common when larger samples are considered.

The brown tree snake (*Boiga irregularis*)

Invasion was more likely in ecosystems that were similar to the one in which the potential invader evolved. Island ecosystems may be more prone to invasion because their species faced few strong competitors and predators, or because their distance from colonizing species populations makes them more likely to have "open" niches. An example of this phenomenon was the decimation of native bird populations on Guam by the invasive brown tree snake. Conversely, invaded ecosystems may lack the natural competitors and predators that check invasives' growth in their native ecosystems.

Invaded ecosystems may have experienced disturbance, typically human-induced. Such a disturbance may give invasive species a chance to establish themselves with less competition from natives less able to adapt to a disturbed ecosystem.

Vectors

Non-native species have many *vectors*, including biogenic vectors, but most invasions are associated with human activity. Natural range extensions are common in many species, but the rate and magnitude of human-mediated extensions in these species tend to be much larger than natural extensions, and humans typically carry specimens greater distances than natural forces.

An early human vector occurred when prehistoric humans introduced the Pacific rat (*Rattus exulans*) to Polynesia.

Chinese mitten crab (*Eriocheir sinensis*)

Vectors include plants or seeds imported for horticulture. The pet trade moves animals across borders, where they can escape and become invasive. Organisms stow away on transport vehicles. Ballast water taken up at sea and released in port by transoceanic vessels is the largest vector for non-native aquatic species invasions. Around the world on the average day, more than 3,000 different species of aquatic life may be transported on these vessels. For example, freshwater zebra mussels, native to the Black, Caspian and Azov seas, probably reached the Great Lakes via ballast water from a transoceanic vessel. Although the zebra mussel invasion was first noted in 1988, and a mitigation plan was successfully implemented shortly thereafter, the plan had (and continued to have as of 2005) a serious flaw or loophole, whereby ships that are loaded with cargo when they reach the Seaway need not be tested, but all the same they transfer ballast 'puddles' between Seaway ports.

The arrival of invasive propagules to a new site is a function of the site's invasibility.

Species have also been introduced intentionally. For example, to feel more "at home," American colonists formed "Acclimation Societies" that repeatedly imported birds that were native to Europe to North America and other distant lands. In 2008, U.S. postal workers in Pennsylvania noticed noises coming from inside a box from Taiwan; the box contained more than two dozen live beetles. Agricultural Research Service entomologists identified them as rhinoceros beetle, hercules beetle, and king stag beetle. Because these species were not native to the U.S., they could have threatened native ecosystems. To prevent exotic species from becoming a problem in the U.S., special handling and permits are required when living materials are shipped from foreign countries. USDA programs such as Smuggling Interdiction and Trade Compliance (SITC) attempt to prevent exotic species outbreaks in America.

Economics plays a major role in exotic species introduction. High demand for the valuable Chinese mitten crab is one explanation for the possible intentional release of the species in foreign waters.

Impacts of Wildfire

Invasive species often exploit disturbances to an ecosystem (wildfires, roads, foot trails) to colonize an area. Large wildfires can sterilize soils, while adding a variety of nutrients. In the resulting free-for-all, formerly entrenched species lose their advantage, leaving more room for invasives. In such circumstances plants that can regenerate from their roots have an advantage. Non-natives with this ability can benefit from a low intensity fire burns that removes surface vegetation, leaving natives that rely on seeds for propagation to find their niches occupied when their seeds finally sprout.

Impact of Wildfire Suppression on Spreading

Wildfires often occur in remote areas, needing fire suppression crews to travel through pristine forest to reach the site. The crews can bring invasive seeds with them. If any of these stowaway seeds become established, a thriving colony of invasives can erupt in as few as six weeks, after which controlling the outbreak can need years of continued attention to prevent further spread. Also, disturbing the soil surface, such as cutting firebreaks, destroys native cover, exposes soil, and can accelerate invasions. In suburban and wildland-urban interface areas, the vegetation clearance and brush removal ordinances of municipalities for defensible space can result in excessive removal of native shrubs and perennials that exposes the soil to more light and less competition for invasive plant species.

Fire suppression vehicles are often major culprits in such outbreaks, as the vehicles are often driven on back roads often overgrown with invasive plant species. The undercarriage of the vehicle becomes a prime vessel of transport. In response, on large fires, washing stations "decontaminate" vehicles before engaging in suppression activities. Large wildfires attract firefighters from remote places, further increasing the potential for seed transport.

Effects

An American alligator attacking a Burmese python in Florida; the Burmese python is an invasive species which is posing a threat to many indigenous species, including the alligator

Ecological

Land clearing and human habitation put significant pressure on local species. Disturbed habitats are prone to invasions that can have adverse effects on local ecosystems, changing ecosystem functions. A species of wetland plant known as ʻaeʻae in Hawaii (the indigenous *Bacopa monnieri*) is regarded as a pest species in artificially manipulated water bird refuges because it quickly covers shallow mudflats established for endangered Hawaiian stilt (*Himantopus mexicanus knudseni*), making these undesirable feeding areas for the birds.

Multiple successive introductions of different non-native species can have interactive effects; the introduction of a second non-native species can enable the first invasive species to flourish. Examples of this are the introductions of the amethyst gem clam (*Gemma gemma*) and the European green crab (*Carcinus maenas*). The gem clam was introduced into California's Bodega Harbor from the East Coast of the United States a century ago. It had been found in small quantities in the harbor but had never displaced the native clam species (*Nutricola* spp.). In the mid-1990s, the introduction of the European green crab, found to prey preferentially on the native clams, resulted in a decline of the native clams and an increase of the introduced clam populations.

In the Waterberg region of South Africa, cattle grazing over the past six centuries has allowed invasive scrub and small trees to displace much of the original grassland, resulting in a massive reduction in forage for native bovids and other grazers. Since the 1970s, large scale efforts have been underway to reduce invasive species; partial success has led to re-establishment of many species that had dwindled or left the region. Examples of these species are giraffe, blue wildebeest, impala, kudu and white rhino.

Invasive species can change the functions of ecosystems. For example, invasive plants can alter the fire regimen (cheatgrass, *Bromus tectorum*), nutrient cycling (smooth cordgrass *Spartina alterniflora*), and hydrology (*Tamarix*) in native ecosystems. Invasive species that are closely related to rare native species have the potential to hybridize with the native species. Harmful effects of hybridization have led to a decline and even extinction of native species. For example, hybridization with introduced cordgrass, *Spartina alterniflora*, threatens the existence of California cordgrass (*Spartina foliosa*) in San Francisco Bay. Invasive species cause competition for native species and because of this 400 of the 958 endangered species under the Endangered Species Act are at risk

Geomorphological

Primary geomorphological effects of invasive plants are bioconstruction and bioprotection. For example, Kudzu Pueraria montana, a vine native to Asia was widely introduced in the southeastern USA in the early 20th century to control soil erosion. While primary effects of invasive animals are bioturbation, bioerosion, and bioconstruction. For example, invasion of Chinese mitten crab Eriocheir sinensis have resulted in higher bioturbation and bioerosion rates.

Economic

Benefits

Non-native species can have benefits. Asian oysters, for example, filter water pollutants better than native oysters. They also grow faster and withstand disease better than natives. Biologists are currently considering releasing this mollusk in the Chesapeake Bay to help restore oyster stocks and remove pollution. A recent study by the Johns Hopkins School of Public Health found the Asian oyster could significantly benefit the bay's deteriorating water quality. Additionally, some species have invaded an area so long ago that they have found their own beneficial niche in the environment. For example, *L. leucozonium*, shown by population genetic analysis to be an invasive species in North America, has become an important pollinator of caneberry as well as cucurbit, apple trees, and blueberry bushes.

Costs

Economic costs from invasive species can be separated into direct costs through production loss in agriculture and forestry, and management costs. Estimated damage and control cost of invasive species in the U.S. alone amount to more than $138 billion annually. Economic losses can also occur through loss of recreational and tourism revenues. When economic costs of invasions are calculated as production loss and management costs, they are low because they do not consider environmental damage; if monetary values were assigned to the extinction of species, loss in biodiversity, and loss of ecosystem services, costs from impacts of invasive species would drastically increase. The following examples from different sectors of the economy demonstrate the impact of biological invasions.

Economic Opportunities

Some invasions offer potential commercial benefits. For instance, silver carp and common carp can be harvested for human food and exported to markets already familiar with the product, or processed into pet foods, or mink feed. Vegetative invasives such as water hyacinth can be turned into fuel by methane digesters.

Invasivorism

Invasive species are flora and fauna whose introduction into a habitat disrupts the native eco-system. In response, Invasivorism is a movement that explores the idea of eating invasive species in order to control, reduce, or eliminate their populations. Chefs from around the world have begun seeking out and using invasive species as alternative ingredients. Miya's of New Haven, Connecticut created the first invasive species menu in the world. Skeptics point out that once a foreign species has entrenched itself in a new place—such as the Indo-Pacific lionfish that has now virtually taken over the waters of the Western Atlantic, Caribbean and Gulf of Mexico—eradication is almost impossible. Critics argue that encouraging consumption might have the unintended effect of spreading harmful species even more widely.

A dish that features whole fried invasive lionfish at Fish Fish of Miami, Florida

Proponents of invasivorism argue that humans have the ability to eat away any species that it has an appetite for, pointing to the many animals which humans have been able to hunt to extinction - such as the Dodo bird, the Caribbean monk seal, and the Passenger pigeon. Proponents of invasivorism also point to the success that Jamaica has had in significantly decreasing the population of lionfish by encouraging the consumption of the fish.

Plant Industry

Weeds reduce yield in agriculture, though they may provide essential nutrients. Some deep-rooted weeds can "mine" nutrients from the subsoil and deposit them on the topsoil, while others provide habitat for beneficial insects or provide foods for pest species. Many weed species are accidental introductions that accompany seeds and imported plant material. Many introduced weeds in pastures compete with native forage plants, threaten young cattle (e.g., leafy spurge, *Euphorbia esula*) or are unpalatable because of thorns and spines (e.g., yellow starthistle). Forage loss from invasive weeds on pastures amounts to nearly US$1 billion in the U.S. alone. A decline in pollinator services and loss of fruit production has been caused by honey bees infected by the invasive varroa mite. Introduced rats (*Rattus rattus* and *R. norvegicus*) have become serious pests on farms, destroying stored grains.

Invasive plant pathogens and insect vectors for plant diseases can also suppress agricultural yields and nursery stock. Citrus greening is a bacterial disease vectored by the invasive Asian citrus psyllid (ACP). Because of the impacts of this disease on citrus crops, citrus is under quarantine and highly regulated in areas where ACP has been found.

Aquaculture

Aquaculture is a very common vector of species introductions – mainly of species with economic potential (e.g., Oreochromis niloticus)

Forestry

Poster asking campers to not move firewood around, avoiding the spread of invasive species.

The unintentional introduction of forest pest species and plant pathogens can change forest ecology and damage the timber industry. Overall, forest ecosystems in the U.S. are widely invaded by exotic pests, plants, and pathogens.

The Asian long-horned beetle (*Anoplophora glabripennis*) was first introduced into the U.S. in 1996, and was expected to infect and damage millions of acres of hardwood trees. As of 2005 thirty million dollars had been spent in attempts to eradicate this pest and protect millions of trees in the affected regions. The woolly adelgid has inflicted damage on old-growth spruce, fir and hemlock forests and damages the Christmas tree industry. And the chestnut blight fungus (*Cryphonectria parasitica*) and Dutch elm disease (*Ophiostoma novo-ulmi*) are two plant pathogens with serious impacts on these two species, and forest health. Garlic mustard, *Alliaria petiolata*, is one of the most problematic invasive plant species in eastern North American forests. The characteristics of garlic mustard are slightly different from those of the surrounding native plants, which results in a highly successful species that is altering the composition and function of the native communities it invades. When garlic mustard invades the understory of a forest, it affects the growth rate of tree seedlings, which is likely to alter forest regeneration of impact forest composition in the future.

Tourism and Recreation

Invasive species can impact outdoor recreation, such as fishing, hunting, hiking, wildlife viewing, and water-based activities. They can damage a wide array of environmental services that are important to recreation, including, but not limited to, water quality and quantity, plant and animal diversity, and species abundance. Eiswerth states, "very little research has been performed to estimate the corresponding economic losses at spatial scales such as regions, states, and watersheds." Eurasian watermilfoil (*Myriophyllum spicatum*) in parts of the US, fill lakes with plants complicating fishing and boating. The very loud call of the introduced common coqui depresses real estate values in affected neighborhoods of Hawaii.

Health

Encroachment of humans into previously remote ecosystems has exposed exotic diseases such as HIV to the wider population. Introduced birds (e.g. pigeons), rodents and insects (e.g. mosquito, flea, louse and tsetse fly pests) can serve as vectors and reservoirs of human afflictions. The introduced Chinese mitten crabs are carriers of Asian lung fluke. Throughout recorded history, epidemics of human diseases, such as malaria, yellow fever, typhus, and bubonic plague, spread via these vectors. A recent example of an introduced disease is the spread of the West Nile virus, which killed humans, birds, mammals, and reptiles. Waterborne disease agents, such as cholera bacteria (*Vibrio cholerae*), and causative agents of harmful algal blooms are often transported via ballast water. Invasive species and accompanying control efforts can have long term public health implications. For instance, pesticides applied to treat a particular pest species could pollute soil and surface water.

Biodiversity

Biotic invasion is considered one of the five top drivers for global biodiversity loss and is increasing because of tourism and globalization. This may be particularly true in inadequately regulated fresh water systems, though quarantines and ballast water rules have improved the situation.

Invasive species may drive local native species to extinction via competitive exclusion, niche displacement, or hybridisation with related native species. Therefore, besides their economic ramifications, alien invasions may result in extensive changes in the structure, composition and global distribution of the biota of sites of introduction, leading ultimately to the homogenisation of the world's fauna and flora and the loss of biodiversity. Nevertheless, it is difficult to unequivocally attribute extinctions to a species invasion, and the few scientific studies that have done so have been with animal taxa. Concern over the impacts of invasive species on biodiversity must therefore consider the actual evidence (either ecological or economic), in relation to the potential risk.

Genetic Pollution

Native species can be threatened with extinction through the process of *genetic pollution*. Genetic pollution is unintentional hybridization and introgression, which leads to homogenization or replacement of local genotypes as a result of either a numerical or fitness advantage of the introduced species. Genetic pollution can operate either through introduction or through habitat modification, bringing previously isolated species into contact. Hybrids resulting from rare species that interbreed with abundant species can swamp the rarer species' gene pool. This is not always apparent from morphological observations alone. Some degree of gene flow is normal, and preserves constellations of genes and genotypes. An example of this is the interbreeding of migrating coyotes with the red wolf, in areas of eastern North Carolina where the red wolf was reintroduced.

Study

Stage	Characteristic
0	Propagules residing in a donor region
I	Traveling
II	Introduced
III	Localized and numerically rare
IVa	Widespread but rare
IVb	Localized but dominant
V	Widespread and dominant

While the study of invasive species can be done within many subfields of biology, the majority of research on invasive organisms has been within the field of ecology and geography where the issue of biological invasions is especially important. Much of the study of invasive species has been influenced by Charles Elton's 1958 book *The Ecology of Invasion by Animals and Plants* which drew upon the limited amount of research done within disparate fields to create a generalized picture of biological invasions. Studies on invasive species remained sparse until the 1990s when research in the field experienced a large amount of growth which continues to this day. This research, which has largely consisted of field observational studies, has disproportionately been concerned with terrestrial plants. The rapid growth of the field has driven a need to standardize the language used to describe invasive species and events. Despite this, little standard terminology exists within the study of invasive species which itself lacks any official designation but is commonly referred to as "Invasion ecology" or more generally "Invasion biology". This lack of standard terminology is a significant problem, and has largely arisen due to the interdisciplinary nature of the field which borrows terms from numerous disciplines such as agriculture, zoology, and pathology, as well as due to studies on invasive species being commonly performed in isolation of one another.

In an attempt to avoid the ambiguous, subjective, and pejorative vocabulary that so often accompanies discussion of invasive species even in scientific papers, Colautti and MacIsaac proposed a new nomenclature system based on biogeography rather than on taxa.

By discarding taxonomy, human health, and economic factors, this model focused only on ecological factors. The model evaluated individual populations rather than entire species. It classified each population based on its success in that environment. This model applied equally to indigenous and to introduced species, and did not automatically categorize successful introductions as harmful.

References

- Shogren, Jason F.; Tschirhart, John (eds.). Protecting Endangered Species in the United States: Biological Needs, Political Realities, Economic Choices. Cambridge University Press. p. 1. ISBN 0521662109.

- Farr, Daniel (2002). Indicator Species. in Encyclopedia of Environmetrics (eds. A H El-Sharaawi and W W Piegorsch), John Wiley & Sons, Ltd. ISBN 978-0-471-89997-6.

- Shrivastava, Rahul (2007). Indicator Species. in Encyclopedia of Environment and Society (ed. Paul Robins), Thousand Oaks, CA: Sage Publications. ISBN 1-4129-2761-7.

- Elton, C.S. (2000) [1958]. The Ecology of Invasions by Animals and Plants. Foreword by Daniel Simberloff. Chicago: University of Chicago Press. p. 196. ISBN 0-226-20638-6.

- Leakey, Richard; Roger Lewin (1999) [1995]. "11 The modern elephant story". The sixth extinction: biodiversity and its survival. London: Phoenix. pp. 216–217. ISBN 1-85799-473-6.

- Lockwood, Julie L.; Martha F. Hoopes; Michael P. Marchetti (2007). Invasion Ecology (PDF). Blackwell Publishing. p. 7. Retrieved 21 January 2014.

- F. Moretzsohn; J.A. Sánchez Chávez; J.W. Tunnell, Jr. (eds.). "Invasive Species". GulfBase: Resource Database for Gulf of Mexico Research. Harte Research Institute for Gulf of Mexico Studies at Texas A&M University-Corpus Christi. Retrieved March 19, 2013.

- Rosmarino, Nicole (2007). "Associated Species : Prairie Dogs are a Keystone Species of the Great Plains". Prairie Dog Coalition. Retrieved 10 November 2013.

- Ducarme, Frédéric; Luque, Gloria M.; Courchamp, Franck (2013). "What are "charismatic species" for conservation biologists ?" (PDF). BioSciences Master Reviews. 1. Retrieved 19 December 2013.

4

Threats to Biodiversity

Human beings are one of the biggest threats to biodiversity. Human activity has led to habitat destruction, extinction of several species, pollution, climate change, overexploitation, human overpopulation etc. This chapter discusses the threats to biodiversity with special reference to these topics. One is able to understand the impact humans have on the environment by their day-to-day activities.

Deforestation and Climate Change

Deforestation is one of the main causes of climate change. It is the second largest anthropogenic source of carbon dioxide to the atmosphere, after fossil fuel combustion. Deforestation and forest degradation contribute to atmospheric greenhouse gas emissions through combustion of forest biomass and decomposition of remaining plant material and soil carbon. It used to account for more than 20% of carbon dioxide emissions, but it's currently somewhere around the 10% mark. By 2008, deforestation was 12% of total CO_2, or 15% if peatlands are included. These proportions are likely to have fallen since given the continued rise of fossil fuel use.

Averaged over all land and ocean surfaces, temperatures warmed roughly 1.53 °F (0.85 °C) between 1880 and 2012, according to the Intergovernmental Panel on Climate Change. In the Northern Hemisphere, 1983 to 2012 were the warmest 30-year period of the last 1400 years.

Effect on Climate Change

Decrease in Biodiversity

A 2007 study conducted by the National Science Foundation found that biodiversity and genetic diversity are codependent—that diversity among species requires diversity within a species, and vice versa. "If any one type is removed from the system, the cycle can break down, and the community becomes dominated by a single species."

Counteracting Climate Change

Reforestation

Reforestation is the natural or intentional restocking of existing forests and woodlands that have been depleted, usually through deforestation. It is the reestablishment of forest cover either naturally or artificially. Similar to the other methods of forestation, reforestation can be very effective because a single tree can absorb as much as 48 pounds of carbon dioxide per year and can sequester 1 ton of carbon dioxide by the time it reaches 40 years old.

Afforestation

Afforestation is the establishment of a forest or stand of trees in an area where there was no forest.

China

Although China has set official goals for reforestation, these goals were set for an 80-year time horizon and were not significantly met by 2008. China is trying to correct these problems with projects such as the Green Wall of China, which aims to replant forests and halt the expansion of the Gobi Desert. A law promulgated in 1981 requires that every school student over the age of 11 plant at least one tree per year. But average success rates, especially in state-sponsored plantings, remains relatively low. And even the properly planted trees have had great difficulty surviving the combined impacts of prolonged droughts, pest infestation and fires. Nonetheless, China currently has the highest afforestation rate of any country or region in the world, with 4.77 million hectares (47,000 square kilometers) of afforestation in 2008.

Japan

The primary goal of afforestation projects in Japan is to develop the forest structure of the nation and to maintain the biodiversity found in the Japanese wilderness. The Japanese temperate rainforest is scattered throughout the Japanese archipelago and is home to many endemic species that are not naturally found anywhere else. As development of the country's caused a decline in forest cover, a reduction in biodiversity was seen in those areas.

Agroforestry

Agroforestry or agro-sylviculture is a land use management system in which trees or shrubs are grown around or among crops or pastureland. It combines agricultural and forestry technologies to create more diverse, productive, profitable, healthy, and sustainable land-use systems.

Projects and Foundations

Arbor Day Foundation

Founded in 1972, the centennial of the first Arbor Day observance in the 19th century, the Foundation has grown to become the largest nonprofit membership organization dedicated to planting trees, with over one million members, supporters, and valued partners. They work on projects focused on planting trees around campuses, low-income communities, and communities that have been affected by natural disasters among other places.

Billion Tree Campaign

The Billion Tree Campaign was launched in 2006 by the United Nations Environment Programme (UNEP) as a response to the challenges of global warming, as well as to a wider array of sustainability challenges, from water supply to biodiversity loss. Its initial target was the planting of one billion trees in 2007. Only one year later in 2008, the campaign's objective was raised to 7 billion trees—a target to be met by the climate change conference that was held in Copenhagen, Denmark in December 2009. Three months before the conference, the 7 billion planted trees mark had been surpassed. In December 2011, after more than 12 billion trees had been planted, UNEP formally handed management of the program over to the not-for-profit Plant-for-the-Planet initiative, based in Munich, Germany.

The Amazon Fund (Brazil)

Considered the largest reserve of biological diversity in the world, the Amazon Basin is also the largest Brazilian biome, taking up almost half the nation's territory. The Amazon Basin corresponds to two fifths of South America's territory. Its area of approximately seven million square kilometers covers the largest hydrographic network on the planet, through which runs about one fifth of the fresh water on the world's surface. Deforestation in the Amazon rainforest is a major cause to climate change due to the decreasing number of trees available to capture increasing carbon dioxide levels in the atmosphere.

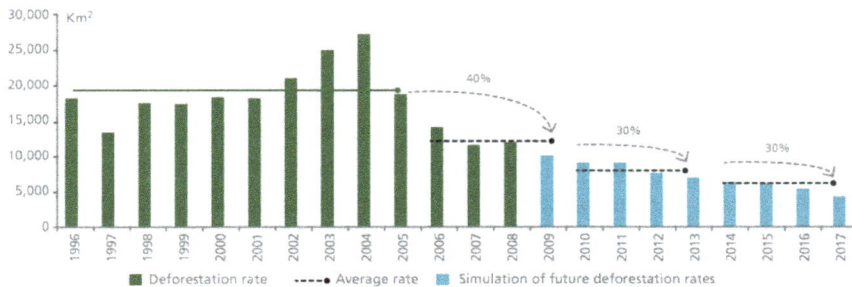

Source: National Plan on Climate Change (PNMC).

Four-year plan to reduce in deforestation in the Amazon

The Amazon Fund is aimed at raising donations for non-reimbursable investments in efforts to prevent, monitor and combat deforestation, as well as to promote the preservation and sustainable use of forests in the Amazon Biome, under the terms of Decree N.º 6,527, dated August 1, 2008. The Amazon Fund supports the following areas: management of public forests and protected areas, environmental control, monitoring and inspection, sustainable forest management, economic activities created with sustainable use of forests, ecological and economic zoning, territorial arrangement and agricultural regulation, preservation and sustainable use of biodiversity, and recovery of deforested areas. Besides those, the Amazon Fund may use up to 20% of its donations to support the development of systems to monitor and control deforestation in other Brazilian biomes and in biomes of other tropical countries.

Climate Change

Climate change is a change in the statistical distribution of weather patterns when that change lasts for an extended period of time (i.e., decades to millions of years). Climate change may refer to a change in average weather conditions, or in the time variation of weather around longer-term average conditions (i.e., more or fewer extreme weather events). Climate change is caused by factors such as biotic processes, variations in solar radiation received by Earth, plate tectonics, and volcanic eruptions. Certain human activities have also been identified as significant causes of recent climate change, often referred to as *global warming*.

Scientists actively work to understand past and future climate by using observations and theoretical models. A climate record—extending deep into the Earth's past—has been assembled, and continues to be built up, based on geological evidence from borehole temperature profiles, cores removed from deep accumulations of ice, floral and faunal records, glacial and periglacial processes, stable-isotope and other analyses of sediment layers, and records of past sea levels. More recent data are provided by the instrumental record. General circulation models, based on the physical sciences, are often used in theoretical approaches to match past climate data, make future projections, and link causes and effects in climate change.

Terminology

The most general definition of *climate change* is a change in the statistical properties (principally its mean and spread) of the climate system when considered over long periods of time, regardless of cause. Accordingly, fluctuations over periods shorter than a few decades, such as El Niño, do not represent climate change.

The term sometimes is used to refer specifically to climate change caused by human activity, as opposed to changes in climate that may have resulted as part of Earth's natural processes. In this sense, especially in the context of environmental policy, the term *climate change* has become synonymous with *anthropogenic global warming*. Within scientific journals, *global warming* refers to surface temperature increases while *climate change* includes global warming and everything else that increasing greenhouse gas levels affect.

Climatic Change Versus Climate Change

In 1966, the World Meteorological Organization (WMO) proposed the term climatic change to encompass all forms of climatic variability on time-scales longer than 10 years, whether the cause was natural or anthropogenic. Change was a given and climatic was used as an adjective to describe this kind of change (as opposed to political or economic change). When it was realized that human activities had a potential to drastically alter the climate, the term climate change replaced climatic change as the dominant term to reflect an anthropogenic cause. Climate change was incorporated in the title of the Intergovernmental Panel on Climate Change (IPCC) and the UN Framework Convention on Climate Change (UNFCCC). Climate change, used as a noun, became an issue rather than the technical description of changing weather.

Causes

On the broadest scale, the rate at which energy is received from the Sun and the rate at which it is lost to space determine the equilibrium temperature and climate of Earth. This energy is distributed around the globe by winds, ocean currents, and other mechanisms to affect the climates of different regions.

Factors that can shape climate are called climate forcings or "forcing mechanisms". These include processes such as variations in solar radiation, variations in the Earth's orbit, variations in the albedo or reflectivity of the continents and oceans, mountain-building and continental drift and changes in greenhouse gas concentrations. There are a variety of climate change feedbacks that can either amplify or diminish the initial forcing. Some parts of the climate system, such as the oceans and ice caps, respond more slowly in reaction to climate forcings, while others respond more quickly. There are also key threshold factors which when exceeded can produce rapid change.

Forcing mechanisms can be either "internal" or "external". Internal forcing mechanisms are natural processes within the climate system itself (e.g., the thermohaline circulation). External forcing mechanisms can be either natural (e.g., changes in solar output) or anthropogenic (e.g., increased emissions of greenhouse gases).

Whether the initial forcing mechanism is internal or external, the response of the climate system might be fast (e.g., a sudden cooling due to airborne volcanic ash reflecting sunlight), slow (e.g. thermal expansion of warming ocean water), or a combination (e.g., sudden loss of albedo in the arctic ocean as sea ice melts, followed by more gradual thermal expansion of the water). Therefore, the climate system can respond abruptly, but the full response to forcing mechanisms might not be fully developed for centuries or even longer.

Internal Forcing Mechanisms

Scientists generally define the five components of earth's climate system to include atmosphere, hydrosphere, cryosphere, lithosphere (restricted to the surface soils, rocks, and sediments), and biosphere. Natural changes in the climate system ("internal forcings") result in internal "climate variability". Examples include the type and distribution of species, and changes in ocean currents.

Ocean Variability

The ocean is a fundamental part of the climate system, some changes in it occurring at longer timescales than in the atmosphere, as it has hundreds of times more mass and thus very high thermal inertia, with effects such as the ocean depths still lagging today in temperature adjustment from effects of the Little Ice Age of past centuries).

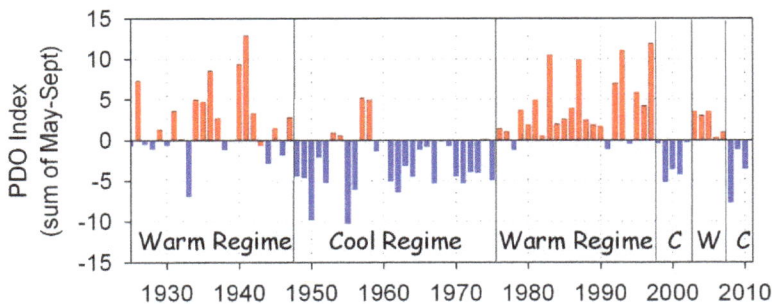

Pacific Decadal Oscillation 1925 to 2010

Short-term fluctuations (years to a few decades) such as the El Niño-Southern Oscillation, the Pacific decadal oscillation, the North Atlantic oscillation, and the Arctic oscillation, represent climate variability rather than climate change. On longer time-scales, alterations to ocean processes such as thermohaline circulation play a key role in redistributing heat by carrying out a very slow and extremely deep movement of water and the long-term redistribution of heat in the world's oceans.

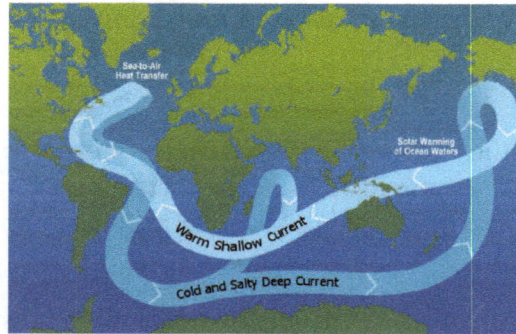

A schematic of modern thermohaline circulation. Tens of millions of years ago, continental-plate movement formed a land-free gap around Antarctica, allowing the formation of the ACC, which keeps warm waters away from Antarctica.

Life

Life affects climate through its role in the carbon and water cycles and through such mechanisms as albedo, evapotranspiration, cloud formation, and weathering. Examples of how life may have affected past climate include:

- glaciation 2.3 billion years ago triggered by the evolution of oxygenic photosynthesis, which depleted the atmosphere of the greenhouse gas carbon dioxide and introduced free oxygen.

- another glaciation 300 million years ago ushered in by long-term burial of decomposition-resistant detritus of vascular land-plants (creating a carbon sink and forming coal)

- termination of the Paleocene-Eocene Thermal Maximum 55 million years ago by flourishing marine phytoplankton

- reversal of global warming 49 million years ago by 800,000 years of arctic azolla blooms

- global cooling over the past 40 million years driven by the expansion of grass-grazer ecosystems

External Forcing Mechanisms

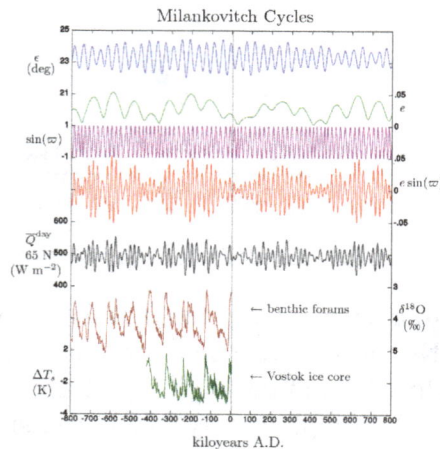

Milankovitch cycles from 800,000 years ago in the past to 800,000 years in the future.

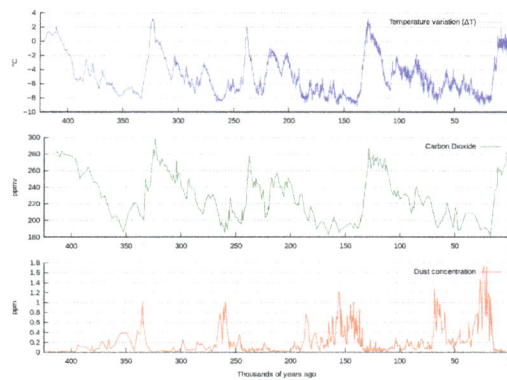

Variations in CO_2, temperature and dust from the Vostok ice core over the last 450,000 years

Orbital Variations

Slight variations in Earth's orbit lead to changes in the seasonal distribution of sunlight reaching the Earth's surface and how it is distributed across the globe. There is very little change to the area-averaged annually averaged sunshine; but there can be strong changes in the geographical and seasonal distribution. The three types of orbital variations are variations in Earth's eccentricity, changes in the tilt angle of Earth's axis of rotation, and precession of Earth's axis. Combined together, these produce Milankovitch cycles which have a large impact on climate and are notable for their correlation to glacial and interglacial periods, their correlation with the advance and retreat of the Sahara, and for their appearance in the stratigraphic record.

The IPCC notes that Milankovitch cycles drove the ice age cycles, CO_2 followed temperature change "with a lag of some hundreds of years," and that as a feedback amplified temperature change. The depths of the ocean have a lag time in changing temperature (thermal inertia on such scale). Upon seawater temperature change, the solubility of CO_2 in the oceans changed, as well as other factors impacting air-sea CO_2 exchange.

Solar Output

The Sun is the predominant source of energy input to the Earth. Other sources include geothermal energy from the Earth's core, and heat from the decay of radioactive compounds. Both long- and short-term variations in solar intensity are known to affect global climate.

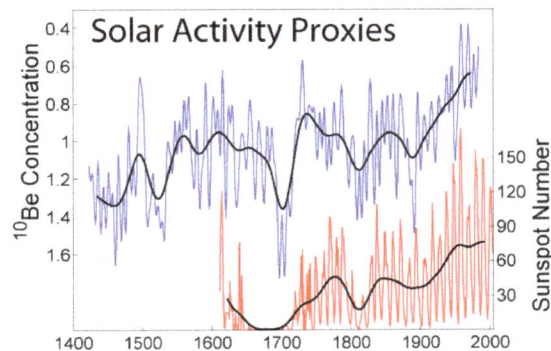

Variations in solar activity during the last several centuries based on observations of sunspots and beryllium isotopes. The period of extraordinarily few sunspots in the late 17th century was the Maunder minimum.

Three to four billion years ago, the Sun emitted only 70% as much power as it does today. If the atmospheric composition had been the same as today, liquid water should not have existed on Earth. However, there is evidence for the presence of water on the early Earth, in the Hadean and Archean eons, leading to what is known as the faint young Sun paradox. Hypothesized solutions to this paradox include a vastly different atmosphere, with much higher concentrations of greenhouse gases than currently exist. Over the following approximately 4 billion years, the energy output of the Sun increased and atmospheric composition changed. The Great Oxygenation Event – oxygenation of the atmosphere around 2.4 billion years ago – was the most notable alteration. Over the next five billion years, the Sun's ultimate death as it becomes a red giant and then a white dwarf will have large effects on climate, with the red giant phase possibly ending any life on Earth that survives until that time.

Solar output also varies on shorter time scales, including the 11-year solar cycle and longer-term modulations. Solar intensity variations possibly as a result of the Wolf, Spörer and Maunder Minimum are considered to have been influential in triggering the Little Ice Age, and some of the warming observed from 1900 to 1950. The cyclical nature of the Sun's energy output is not yet fully understood; it differs from the very slow change that is happening within the Sun as it ages and evolves. Research indicates that solar variability has had effects including the Maunder minimum from 1645 to 1715 A.D., part of the Little Ice Age from 1550 to 1850 A.D. that was marked by relative cooling and greater glacier extent than the centuries before and afterward. Some studies point toward solar radiation increases from cyclical sunspot activity affecting global warming, and climate may be influenced by the sum of all effects (solar variation, anthropogenic radiative forcings, etc.).

Interestingly, a 2010 study *suggests*, "that the effects of solar variability on temperature throughout the atmosphere may be contrary to current expectations."

In an Aug 2011 Press Release, CERN announced the publication in the Nature journal the initial results from its CLOUD experiment. The results indicate that ionisation from cosmic rays significantly enhances aerosol formation in the presence of sulfuric acid and water, but in the lower atmosphere where ammonia is also required, this is insufficient to account for aerosol formation and additional trace vapours must be involved. The next step is to find more about these trace vapours, including whether they are of natural or human origin.

Volcanism

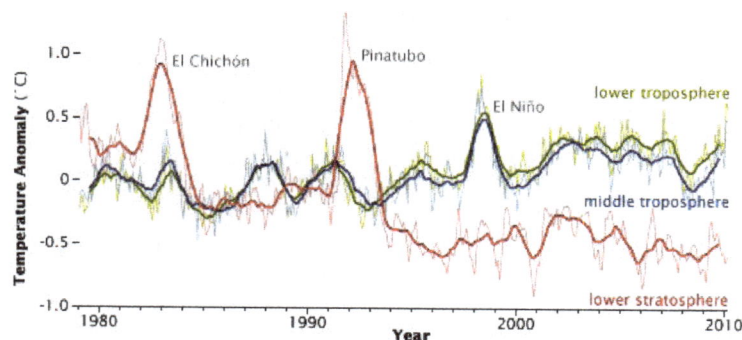

In atmospheric temperature from 1979 to 2010, determined by MSU NASA satellites, effects appear from aerosols released by major volcanic eruptions (El Chichón and Pinatubo). El Niño is a separate event, from ocean variability.

The eruptions considered to be large enough to affect the Earth's climate on a scale of more than 1 year are the ones that inject over 100,000 tons of SO_2 into the stratosphere. This is due to the optical properties of SO_2 and sulfate aerosols, which strongly absorb or scatter solar radiation, creating a global layer of sulfuric acid haze. On average, such eruptions occur several times per century, and cause cooling (by partially blocking the transmission of solar radiation to the Earth's surface) for a period of a few years.

The eruption of Mount Pinatubo in 1991, the second largest terrestrial eruption of the 20th century, affected the climate substantially, subsequently global temperatures decreased by about 0.5 °C (0.9 °F) for up to three years. Thus, the cooling over large parts of the Earth reduced surface temperatures in 1991-93, the equivalent to a reduction in net radiation of 4 watts per square meter. The Mount Tambora eruption in 1815 caused the Year Without a Summer. Much larger eruptions, known as large igneous provinces, occur only a few times every fifty - hundred million years - through flood basalt, and caused in Earth past global warming and mass extinctions.

Small eruptions, with injections of less than 0.1 Mt of sulfur dioxide into the stratosphere, impact the atmosphere only subtly, as temperature changes are comparable with natural variability. However, because smaller eruptions occur at a much higher frequency, they too have a significant impact on Earth's atmosphere.

Seismic monitoring maps current and future trends in volcanic activities, and tries to develop early warning systems. In climate modelling the aim is to study the physical mechanisms and feedbacks of volcanic forcing.

Volcanoes are also part of the extended carbon cycle. Over very long (geological) time periods, they release carbon dioxide from the Earth's crust and mantle, counteracting the uptake by sedimentary rocks and other geological carbon dioxide sinks. The US Geological Survey estimates are that volcanic emissions are at a much lower level than the effects of current human activities, which generate 100–300 times the amount of carbon dioxide emitted by volcanoes. A review of published studies indicates that annual volcanic emissions of carbon dioxide, including amounts released from mid-ocean ridges, volcanic arcs, and hot spot volcanoes, are only the equivalent of 3 to 5 days of human caused output. The annual amount put out by human activities may be greater than the amount released by supererruptions, the most recent of which was the Toba eruption in Indonesia 74,000 years ago.

Although volcanoes are technically part of the lithosphere, which itself is part of the climate system, the IPCC explicitly defines volcanism as an external forcing agent.

Plate Tectonics

Over the course of millions of years, the motion of tectonic plates reconfigures global land and ocean areas and generates topography. This can affect both global and local patterns of climate and atmosphere-ocean circulation.

The position of the continents determines the geometry of the oceans and therefore influences patterns of ocean circulation. The locations of the seas are important in controlling the transfer of heat and moisture across the globe, and therefore, in determining global climate. A recent example of tectonic control on ocean circulation is the formation of the Isthmus of Panama about

5 million years ago, which shut off direct mixing between the Atlantic and Pacific Oceans. This strongly affected the ocean dynamics of what is now the Gulf Stream and may have led to Northern Hemisphere ice cover. During the Carboniferous period, about 300 to 360 million years ago, plate tectonics may have triggered large-scale storage of carbon and increased glaciation. Geologic evidence points to a "megamonsoonal" circulation pattern during the time of the supercontinent Pangaea, and climate modeling suggests that the existence of the supercontinent was conducive to the establishment of monsoons.

The size of continents is also important. Because of the stabilizing effect of the oceans on temperature, yearly temperature variations are generally lower in coastal areas than they are inland. A larger supercontinent will therefore have more area in which climate is strongly seasonal than will several smaller continents or islands.

Human Influences

In the context of climate variation, anthropogenic factors are human activities which affect the climate. The scientific consensus on climate change is "that climate is changing and that these changes are in large part caused by human activities," and it "is largely irreversible."

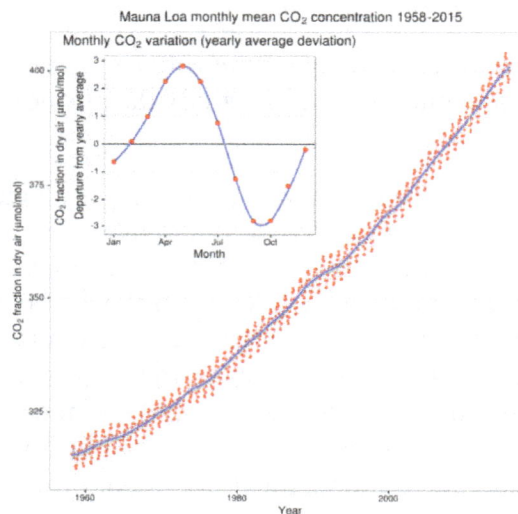

Increase in atmospheric CO_2 levels

"Science has made enormous inroads in understanding climate change and its causes, and is beginning to help develop a strong understanding of current and potential impacts that will affect people today and in coming decades. This understanding is crucial because it allows decision makers to place climate change in the context of other large challenges facing the nation and the world. There are still some uncertainties, and there always will be in understanding a complex system like Earth's climate. Nevertheless, there is a strong, credible body of evidence, based on multiple lines of research, documenting that climate is changing and that these changes are in large part caused by human activities. While much remains to be learned, the core phenomenon, scientific questions, and hypotheses have been examined thoroughly and have stood firm in the face of serious scientific debate and careful evaluation of alternative explanations."

— *United States National Research Council, Advancing the Science of Climate Change*

Of most concern in these anthropogenic factors is the increase in CO_2 levels due to emissions from fossil fuel combustion, followed by aerosols (particulate matter in the atmosphere) and the CO_2 released by cement manufacture. Other factors, including land use, ozone depletion, animal agriculture and deforestation, are also of concern in the roles they play – both separately and in conjunction with other factors – in affecting climate, microclimate, and measures of climate variables.

Physical Evidence

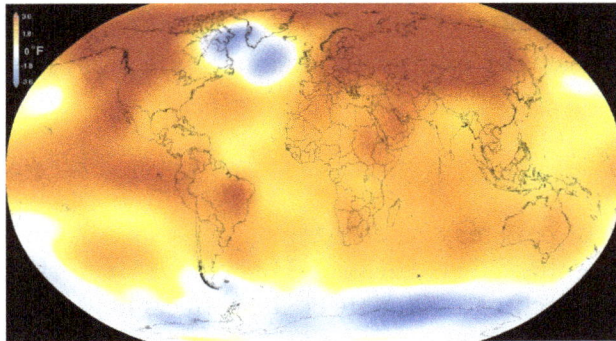

2015 – Warmest Global Year on Record (since 1880) – Colors indicate temperature anomalies (NASA/NOAA; 20 January 2016).

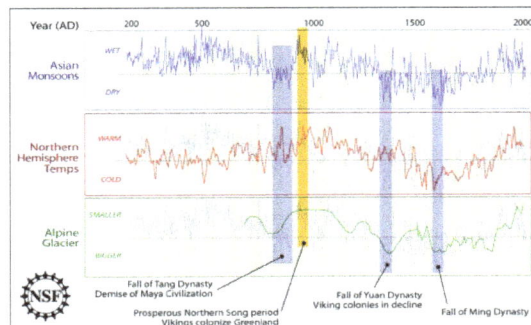

Comparisons between Asian Monsoons from 200 AD to 2000 AD (staying in the background on other plots), Northern Hemisphere temperature, Alpine glacier extent (vertically inverted as marked), and human history as noted by the U.S. NSF.

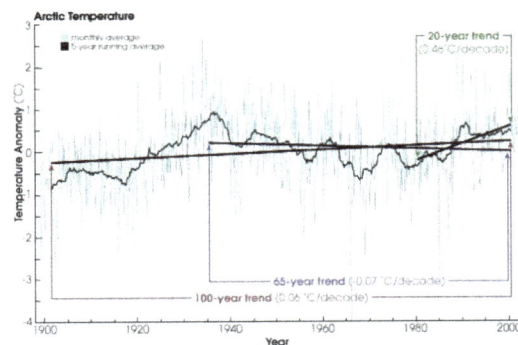

Arctic temperature anomalies over a 100-year period as estimated by NASA. Typical high monthly variance can be seen, while longer-term averages highlight trends.

Evidence for climatic change is taken from a variety of sources that can be used to reconstruct past climates. Reasonably complete global records of surface temperature are available beginning from the mid-late 19th century. For earlier periods, most of the evidence is indirect—climatic changes are inferred from changes in proxies, indicators that reflect climate, such as vegetation, ice cores, dendrochronology, sea level change, and glacial geology.

Temperature Measurements and Proxies

The instrumental temperature record from surface stations was supplemented by radiosonde balloons, extensive atmospheric monitoring by the mid-20th century, and, from the 1970s on, with global satellite data as well. The $^{18}O/^{16}O$ ratio in calcite and ice core samples used to deduce ocean temperature in the distant past is an example of a temperature proxy method, as are other climate metrics noted in subsequent categories.

Historical and Archaeological Evidence

Climate change in the recent past may be detected by corresponding changes in settlement and agricultural patterns. Archaeological evidence, oral history and historical documents can offer insights into past changes in the climate. Climate change effects have been linked to the collapse of various civilizations.

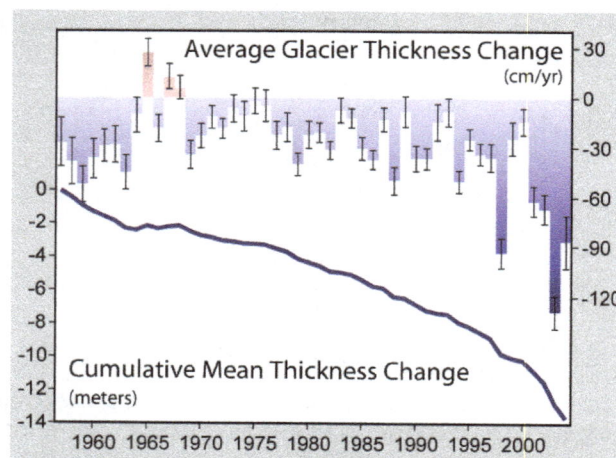

Decline in thickness of glaciers worldwide over the past half-century

Glaciers

Glaciers are considered among the most sensitive indicators of climate change. Their size is determined by a mass balance between snow input and melt output. As temperatures warm, glaciers retreat unless snow precipitation increases to make up for the additional melt; the converse is also true.

Glaciers grow and shrink due both to natural variability and external forcings. Variability in temperature, precipitation, and englacial and subglacial hydrology can strongly determine the evolution of a glacier in a particular season. Therefore, one must average over a decadal or longer time-scale and/or over many individual glaciers to smooth out the local short-term variability and obtain a glacier history that is related to climate.

A world glacier inventory has been compiled since the 1970s, initially based mainly on aerial photographs and maps but now relying more on satellites. This compilation tracks more than 100,000 glaciers covering a total area of approximately 240,000 km², and preliminary estimates indicate that the remaining ice cover is around 445,000 km². The World Glacier Monitoring Service collects data annually on glacier retreat and glacier mass balance. From this data, glaciers worldwide have been found to be shrinking significantly, with strong glacier retreats in the 1940s, stable or growing conditions during the 1920s and 1970s, and again retreating from the mid-1980s to present.

The most significant climate processes since the middle to late Pliocene (approximately 3 million years ago) are the glacial and interglacial cycles. The present interglacial period (the Holocene) has lasted about 11,700 years. Shaped by orbital variations, responses such as the rise and fall of continental ice sheets and significant sea-level changes helped create the climate. Other changes, including Heinrich events, Dansgaard–Oeschger events and the Younger Dryas, however, illustrate how glacial variations may also influence climate without the orbital forcing.

Glaciers leave behind moraines that contain a wealth of material—including organic matter, quartz, and potassium that may be dated—recording the periods in which a glacier advanced and retreated. Similarly, by tephrochronological techniques, the lack of glacier cover can be identified by the presence of soil or volcanic tephra horizons whose date of deposit may also be ascertained.

Arctic Sea Ice Loss

The decline in Arctic sea ice, both in extent and thickness, over the last several decades is further evidence for rapid climate change. Sea ice is frozen seawater that floats on the ocean surface. It covers millions of square miles in the polar regions, varying with the seasons. In the Arctic, some sea ice remains year after year, whereas almost all Southern Ocean or Antarctic sea ice melts away and reforms annually. Satellite observations show that Arctic sea ice is now declining at a rate of 13.3 percent per decade, relative to the 1981 to 2010 average.

This video summarizes how climate change, associated with increased carbon dioxide levels, has affected plant growth.

Vegetation

A change in the type, distribution and coverage of vegetation may occur given a change in the climate. Some changes in climate may result in increased precipitation and warmth, resulting in improved plant growth and the subsequent sequestration of airborne CO_2. A gradual increase in warmth in a region will lead to earlier flowering and fruiting times, driving a change in the timing of life cycles of dependent organisms. Conversely, cold will cause plant bio-cycles to lag. Larger, faster or more radical changes, however, may result in vegetation stress, rapid plant loss and desertification in certain circumstances. An example of this occurred during the Carboniferous Rainforest Collapse (CRC), an extinction event 300 million years ago. At this time vast rainforests covered the equatorial region of Europe and America. Climate change devastated these tropical rainforests, abruptly fragmenting the habitat into isolated 'islands' and causing the extinction of many plant and animal species.

Pollen Analysis

Palynology is the study of contemporary and fossil palynomorphs, including pollen. Palynology is used to infer the geographical distribution of plant species, which vary under different climate conditions. Different groups of plants have pollen with distinctive shapes and surface textures, and since the outer surface of pollen is composed of a very resilient material, they resist decay. Changes in the type of pollen found in different layers of sediment in lakes, bogs, or river deltas indicate changes in plant communities. These changes are often a sign of a changing climate. As an example, palynological studies have been used to track changing vegetation patterns throughout the Quaternary glaciations and especially since the last glacial maximum.

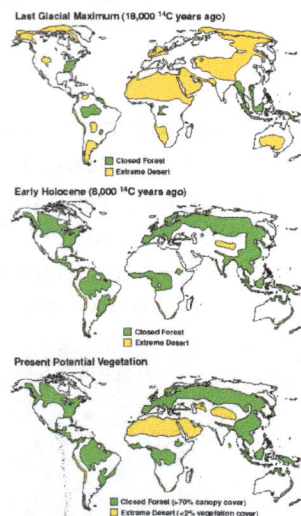

Top: Arid ice age climate
Middle: Atlantic Period, warm and wet
Bottom: Potential vegetation in climate now if not for human effects like agriculture.

Cloud Cover and Precipitation

Past precipitation can be estimated in the modern era with the global network of precipitation gauges. Surface coverage over oceans and remote areas is relatively sparse, but, reducing reliance on interpolation, satellite clouds and precipitation data has been available since the 1970s. Quantification of climatological variation of precipitation in prior centuries and epochs is less complete but approximated using proxies such as marine sediments, ice cores, cave stalagmites, and tree rings. In July 2016 scientists published evidence of increased cloud cover over polar regions, as predicted by climate models.

Climatological temperatures substantially affect cloud cover and precipitation. For instance, during the Last Glacial Maximum of 18,000 years ago, thermal-driven evaporation from the oceans onto continental landmasses was low, causing large areas of extreme desert, including polar deserts (cold but with low rates of cloud cover and precipitation). In contrast, the world's climate was cloudier and wetter than today near the start of the warm Atlantic Period of 8000 years ago.

Estimated global land precipitation increased by approximately 2% over the course of the 20th century, though the calculated trend varies if different time endpoints are chosen, complicated by ENSO and other oscillations, including greater global land cloud cover precipitation in the 1950s and 1970s than the later 1980s and 1990s despite the positive trend over the century overall. Similar slight overall increase in global river runoff and in average soil moisture has been perceived.

Dendroclimatology

Dendroclimatology is the analysis of tree ring growth patterns to determine past climate variations. Wide and thick rings indicate a fertile, well-watered growing period, whilst thin, narrow rings indicate a time of lower rainfall and less-than-ideal growing conditions.

Ice Cores

Analysis of ice in a core drilled from an ice sheet such as the Antarctic ice sheet, can be used to show a link between temperature and global sea level variations. The air trapped in bubbles in the ice can also reveal the CO_2 variations of the atmosphere from the distant past, well before modern environmental influences. The study of these ice cores has been a significant indicator of the changes in CO_2 over many millennia, and continues to provide valuable information about the differences between ancient and modern atmospheric conditions.

Animals

Remains of beetles are common in freshwater and land sediments. Different species of beetles tend to be found under different climatic conditions. Given the extensive lineage of beetles whose genetic makeup has not altered significantly over the millennia, knowledge of the present climatic range of the different species, and the age of the sediments in which remains are found, past climatic conditions may be inferred.

Similarly, the historical abundance of various fish species has been found to have a substantial relationship with observed climatic conditions. Changes in the primary productivity of autotrophs in the oceans can affect marine food webs.

Sea Level Change

Global sea level change for much of the last century has generally been estimated using tide gauge measurements collated over long periods of time to give a long-term average. More recently, altimeter measurements — in combination with accurately determined satellite orbits — have provided an improved measurement of global sea level change. To measure sea levels prior to instrumental measurements, scientists have dated coral reefs that grow near the surface of the ocean, coastal sediments, marine terraces, ooids in limestones, and nearshore archaeological remains. The predominant dating methods used are uranium series and radiocarbon, with cosmogenic radionuclides being sometimes used to date terraces that have experienced relative sea level fall. In the early Pliocene, global temperatures were 1–2 °C warmer than the present temperature, yet sea level was 15–25 meters higher than today.

Deforestation

Deforestation, clearance or clearing is the removal of a forest or stand of trees where the land is thereafter converted to a non-forest use. Examples of deforestation include conversion of forest-land to farms, ranches, or urban use. Tropical rainforests is where the most concentrated defor-estation occurs. About 30% of Earth's land surface is covered by forests.

Satellite photograph of deforestation in progress in the Tierras Bajas project in eastern Bolivia.

In temperate mesic climates, natural regeneration of forest stands often will not occur in the ab-sence of disturbance, whether natural or anthropogenic. Furthermore, biodiversity after regener-ation harvest often mimics that found after natural disturbance, including biodiversity loss after naturally occurring rainforest destruction.

Deforestation occurs for multiple reasons: trees are cut down to be used or sold as fuel (some-times in the form of charcoal) or timber, while cleared land is used as pasture for livestock and plantation. The removal of trees without sufficient reforestation has resulted in damage to habitat, biodiversity loss and aridity. It has adverse impacts on biosequestration of atmospheric carbon dioxide. Deforestation has also been used in war to deprive the enemy of cover for its forces and also vital resources. Modern examples of this were the use of Agent Orange by the British military in Malaya during the Malayan Emergency and the United States military in Vietnam during the Vietnam War. As of 2005, net deforestation rates have ceased to increase in countries with a per capita GDP of at least US$4,600. Deforested regions typically incur significant adverse soil erosion and frequently degrade into wasteland.

Disregard of ascribed value, lax forest management and deficient environmental laws are some of the factors that allow deforestation to occur on a large scale. In many countries, deforestation, both naturally occurring and human induced, is an ongoing issue. Deforestation causes extinction, changes to climatic conditions, desertification, and displacement of populations as observed by current conditions and in the past through the fossil record. More than half of all plant and land animal species in the world live in tropical forests.

Between 2000 and 2012, 2.3 million square kilometres (890,000 square miles) of forests around the earth were cut down. As a result of deforestation, only 6.2 million square kilometres (2.4 million square miles) remain of the original 16 million square kilometres (6 million square miles) of forest that formerly covered the earth.

Causes

According to the United Nations Framework Convention on Climate Change (UNFCCC) secretariat, the overwhelming direct cause of deforestation is agriculture. Subsistence farming is responsible for 48% of deforestation; commercial agriculture is responsible for 32% of deforestation; logging is responsible for 14% of deforestation and fuel wood removals make up 5% of deforestation.

Experts do not agree on whether industrial logging is an important contributor to global deforestation. Some argue that poor people are more likely to clear forest because they have no alternatives, others that the poor lack the ability to pay for the materials and labour needed to clear forest. One study found that population increases due to high fertility rates were a primary driver of tropical deforestation in only 8% of cases.

Other causes of contemporary deforestation may include corruption of government institutions, the inequitable distribution of wealth and power, population growth and overpopulation, and urbanization. Globalization is often viewed as another root cause of deforestation, though there are cases in which the impacts of globalization (new flows of labor, capital, commodities, and ideas) have promoted localized forest recovery.

The last batch of sawnwood from the peat forest in Indragiri Hulu, Sumatra, Indonesia. Deforestation for oil palm plantation.

In 2000 the United Nations Food and Agriculture Organization (FAO) found that "the role of population dynamics in a local setting may vary from decisive to negligible," and that deforestation can result from "a combination of population pressure and stagnating economic, social and technological conditions."

The degradation of forest ecosystems has also been traced to economic incentives that make forest conversion appear more profitable than forest conservation. Many important forest functions have no markets, and hence, no economic value that is readily apparent to the forests' owners or the communities that rely on forests for their well-being. From the perspective of the developing world, the benefits of forest as carbon sinks or biodiversity reserves go primarily to richer developed nations and there is insufficient compensation for these services. Developing countries feel that some countries in the developed world, such as the United States of America, cut down their forests centuries ago and benefited greatly from this deforestation, and that it is hypocritical to deny developing countries the same opportunities: that the poor shouldn't have to bear the cost of preservation when the rich created the problem.

Some commentators have noted a shift in the drivers of deforestation over the past 30 years. Whereas deforestation was primarily driven by subsistence activities and government-sponsored development projects like transmigration in countries like Indonesia and colonization in Latin America, India, Java, and so on, during late 19th century and the earlier half of the 20th century. By the 1990s the majority of deforestation was caused by industrial factors, including extractive industries, large-scale cattle ranching, and extensive agriculture.

Environmental Problems

Atmospheric

Deforestation is ongoing and is shaping climate and geography.

Illegal slash and burn practice in Madagascar, 2010

Deforestation is a contributor to global warming, and is often cited as one of the major causes of the enhanced greenhouse effect. Tropical deforestation is responsible for approximately 20% of world greenhouse gas emissions. According to the Intergovernmental Panel on Climate Change deforestation, mainly in tropical areas, could account for up to one-third of total anthropogenic carbon dioxide emissions. But recent calculations suggest that carbon dioxide emissions from deforestation and forest degradation (excluding peatland emissions) contribute about 12% of total anthropogenic carbon dioxide emissions with a range from 6 to 17%. Deforestation causes carbon dioxide to linger in the atmosphere. As carbon dioxide accrues, it produces a layer in the atmosphere that traps radiation from the sun. The radiation converts to heat which causes global warming, which is better known as the greenhouse effect. Plants remove carbon in the form of carbon dioxide from the atmosphere during the process of photosynthesis, but release some carbon dioxide back into the atmosphere during normal respiration. Only when actively growing can a tree or forest remove carbon, by storing it in plant tissues. Both the decay and burning of wood releases much of this stored carbon back to the atmosphere. In order for forests to take up carbon, there must be a net accumulation of wood. One way is for the wood to be harvested and turned into long-lived products, with new young trees replacing them. Deforestation may also cause carbon stores held in soil to be released. Forests can be either sinks or sources depending upon environmental circumstances. Mature forests alternate between being net sinks and net sources of carbon dioxide

In deforested areas, the land heats up faster and reaches a higher temperature, leading to localized upward motions that enhance the formation of clouds and ultimately produce more rainfall. However, according to the Geophysical Fluid Dynamics Laboratory, the models used to investigate remote responses to tropical deforestation showed a broad but mild temperature increase all through the tropical atmosphere. The model predicted <0.2 °C warming for upper air at 700 mb and 500 mb. However, the model shows no significant changes in other areas besides the Tropics. Though the model showed no significant changes to the climate in areas other than the Tropics, this may not be the case since the model has possible errors and the results are never absolutely definite.

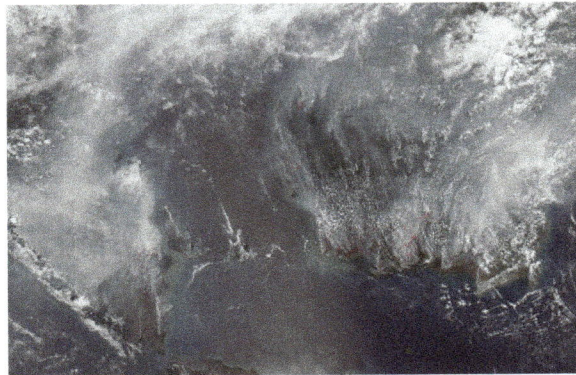

Fires on Borneo and Sumatra, 2006. People use slash-and-burn deforestation to clear land for agriculture.

Reducing emissions from deforestation and forest degradation (REDD) in developing countries has emerged as a new potential to complement ongoing climate policies. The idea consists in providing financial compensations for the reduction of greenhouse gas (GHG) emissions from deforestation and forest degradation".

Rainforests are widely believed by laymen to contribute a significant amount of the world's oxygen, although it is now accepted by scientists that rainforests contribute little net oxygen to the atmosphere and deforestation has only a minor effect on atmospheric oxygen levels. However, the incineration and burning of forest plants to clear land releases large amounts of CO_2, which contributes to global warming. Scientists also state that tropical deforestation releases 1.5 billion tons of carbon each year into the atmosphere.

Hydrological

The water cycle is also affected by deforestation. Trees extract groundwater through their roots and release it into the atmosphere. When part of a forest is removed, the trees no longer transpire this water, resulting in a much drier climate. Deforestation reduces the content of water in the soil and groundwater as well as atmospheric moisture. The dry soil leads to lower water intake for the trees to extract. Deforestation reduces soil cohesion, so that erosion, flooding and landslides ensue.

Shrinking forest cover lessens the landscape's capacity to intercept, retain and transpire precipitation. Instead of trapping precipitation, which then percolates to groundwater systems, deforested areas become sources of surface water runoff, which moves much faster than subsurface flows. That quicker transport of surface water can translate into flash flooding and more localized floods than would occur with the forest cover. Deforestation also contributes to decreased evapotranspiration, which lessens atmospheric moisture which in some cases affects precipitation levels downwind from the deforested area, as water is not recycled to downwind forests, but is lost in runoff and returns directly to the oceans. According to one study, in deforested north and northwest China, the average annual precipitation decreased by one third between the 1950s and the 1980s.

Trees, and plants in general, affect the water cycle significantly:

- their canopies intercept a proportion of precipitation, which is then evaporated back to the atmosphere (canopy interception);

- their litter, stems and trunks slow down surface runoff;

- their roots create macropores – large conduits – in the soil that increase infiltration of water;

- they contribute to terrestrial evaporation and reduce soil moisture via transpiration;

- their litter and other organic residue change soil properties that affect the capacity of soil to store water.

- their leaves control the humidity of the atmosphere by transpiring. 99% of the water absorbed by the roots moves up to the leaves and is transpired.

As a result, the presence or absence of trees can change the quantity of water on the surface, in the soil or groundwater, or in the atmosphere. This in turn changes erosion rates and the availability of water for either ecosystem functions or human services.

The forest may have little impact on flooding in the case of large rainfall events, which overwhelm the storage capacity of forest soil if the soils are at or close to saturation.

Tropical rainforests produce about 30% of our planet's fresh water.

Soil

Deforestation for the use of clay in the Brazilian city of Rio de Janeiro. The hill depicted is Morro da Covanca, in Jacarepaguá

Undisturbed forests have a very low rate of soil loss (erosion), approximately 2 metric tons per square kilometer (6 short tons per square mile). Deforestation generally increases rates of soil loss, by increasing the amount of runoff and reducing the protection of the soil from tree litter. This can be an advantage in excessively leached tropical rain forest soils. Forestry operations themselves also increase erosion through the development of (forest) roads and the use of mechanized equipment.

China's Loess Plateau was cleared of forest millennia ago. Since then it has been eroding, creating dramatic incised valleys, and providing the sediment that gives the Yellow River its yellow color and that causes the flooding of the river in the lower reaches (hence the river's nickname 'China's sorrow').

Removal of trees does not always increase erosion rates. In certain regions of southwest US, shrubs and trees have been encroaching on grassland. The trees themselves enhance the loss of grass between tree canopies. The bare intercanopy areas become highly erodible. The US Forest Service, in Bandelier National Monument for example, is studying how to restore the former ecosystem, and reduce erosion, by removing the trees.

Tree roots bind soil together, and if the soil is sufficiently shallow they act to keep the soil in place by also binding with underlying bedrock. Tree removal on steep slopes with shallow soil thus increases the risk of landslides, which can threaten people living nearby.

Biodiversity

Deforestation on a human scale results in decline in biodiversity, and on a natural global scale is known to cause the extinction of many species. The removal or destruction of areas of forest cover has resulted in a degraded environment with reduced biodiversity. Forests support biodiversity, providing habitat for wildlife; moreover, forests foster medicinal conservation. With forest biotopes being irreplaceable source of new drugs (such as taxol), deforestation can destroy genetic variations (such as crop resistance) irretrievably.

Illegal logging in Madagascar. In 2009, the vast majority of the illegally obtained rosewood was exported to China.

Since the tropical rainforests are the most diverse ecosystems on Earth and about 80% of the world's known biodiversity could be found in tropical rainforests, removal or destruction of significant areas of forest cover has resulted in a degraded environment with reduced biodiversity. A study in Rondônia, Brazil, has shown that deforestation also removes the microbial community which is involved in the recycling of nutrients, the production of clean water and the removal of pollutants.

It has been estimated that we are losing 137 plant, animal and insect species every single day due to rainforest deforestation, which equates to 50,000 species a year. Others state that tropical rainforest deforestation is contributing to the ongoing Holocene mass extinction. The known extinction rates from deforestation rates are very low, approximately 1 species per year from mammals and birds which extrapolates to approximately 23,000 species per year for all species. Predictions have been made that more than 40% of the animal and plant species in Southeast Asia could be wiped out in the 21st century. Such predictions were called into question by 1995 data that show that within regions of Southeast Asia much of the original forest has been converted to monospecific plantations, but that potentially endangered species are few and tree flora remains widespread and stable.

Scientific understanding of the process of extinction is insufficient to accurately make predictions about the impact of deforestation on biodiversity. Most predictions of forestry related biodiversity loss are based on species-area models, with an underlying assumption that as the forest declines species diversity will decline similarly. However, many such models have been proven to be wrong and loss of habitat does not necessarily lead to large scale loss of species. Species-area models are known to overpredict the number of species known to be threatened in areas where actual deforestation is ongoing, and greatly overpredict the number of threatened species that are widespread.

A recent study of the Brazilian Amazon predicts that despite a lack of extinctions thus far, up to 90 percent of predicted extinctions will finally occur in the next 40 years.

Economic Impact

Damage to forests and other aspects of nature could halve living standards for the world's poor and reduce global GDP by about 7% by 2050, a report concluded at the Convention on Biological Diversity (CBD) meeting in Bonn. Historically, utilization of forest products, including timber and fuel wood, has played a key role in human societies, comparable to the roles of water and cultivable land. Today, developed countries continue to utilize timber for building houses, and wood pulp for paper. In developing countries almost three billion people rely on wood for heating and cooking.

The forest products industry is a large part of the economy in both developed and developing countries. Short-term economic gains made by conversion of forest to agriculture, or over-exploitation of wood products, typically leads to loss of long-term income and long-term biological productivity. West Africa, Madagascar, Southeast Asia and many other regions have experienced lower revenue because of declining timber harvests. Illegal logging causes billions of dollars of losses to national economies annually.

The new procedures to get amounts of wood are causing more harm to the economy and overpower the amount of money spent by people employed in logging. According to a study, "in most areas studied, the various ventures that prompted deforestation rarely generated more than US$5 for every ton of carbon they released and frequently returned far less than US$1". The price on the European market for an offset tied to a one-ton reduction in carbon is 23 euro (about US$35).

Rapidly growing economies also have an effect on deforestation. Most pressure will come from the world's developing countries, which have the fastest-growing populations and most rapid economic (industrial) growth. In 1995, economic growth in developing countries reached nearly 6%, compared with the 2% growth rate for developed countries." As our human population grows, new homes, communities, and expansions of cities will occur. Connecting all of the new expansions will be roads, a very important part in our daily life. Rural roads promote economic development but also facilitate deforestation. About 90% of the deforestation has occurred within 100 km of roads in most parts of the Amazon.

The European Union is one of the largest importer of products made from illegal deforestation.

Forest Transition Theory

The forest area change may follow a pattern suggested by the forest transition (FT) theory, whereby at early stages in its development a country is characterized by high forest cover and low deforestation rates (HFLD countries).

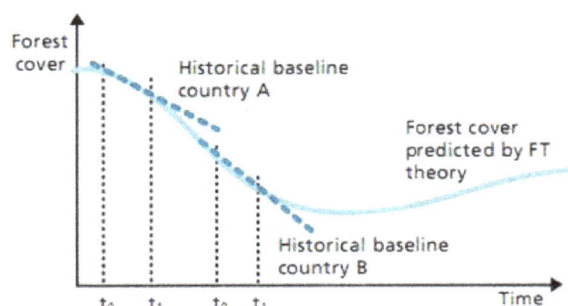

Source: Angelsen 2008.

The forest transition and historical baselines.

Then deforestation rates accelerate (HFHD, high forest cover – high deforestation rate), and forest cover is reduced (LFHD, low forest cover – high deforestation rate), before the deforestation rate slows (LFLD, low forest cover – low deforestation rate), after which forest cover stabilizes and eventually starts recovering. FT is not a "law of nature," and the pattern is influenced by national context (for example, human population density, stage of development, structure of the economy), global economic forces, and government policies. A country may reach very low levels of forest cover before it stabilizes, or it might through good policies be able to "bridge" the forest transition.

FT depicts a broad trend, and an extrapolation of historical rates therefore tends to underestimate future BAU deforestation for counties at the early stages in the transition (HFLD), while it tends to overestimate BAU deforestation for countries at the later stages (LFHD and LFLD).

Countries with high forest cover can be expected to be at early stages of the FT. GDP per capita captures the stage in a country's economic development, which is linked to the pattern of natural resource use, including forests. The choice of forest cover and GDP per capita also fits well with the two key scenarios in the FT:

(i) a forest scarcity path, where forest scarcity triggers forces (for example, higher prices of forest products) that lead to forest cover stabilization; and

(ii) an economic development path, where new and better off-farm employment opportunities associated with economic growth (= increasing GDP per capita) reduce profitability of frontier agriculture and slows deforestation.

Historical Causes

Prehistory

The Carboniferous Rainforest Collapse was an event that occurred 300 million years ago. Climate change devastated tropical rainforests causing the extinction of many plant and animal species. The change was abrupt, specifically, at this time climate became cooler and drier, conditions that are not favourable to the growth of rainforests and much of the biodiversity within them. Rainforests were fragmented forming shrinking 'islands' further and further apart. This sudden collapse affected several large groups, effects on amphibians were particularly devastating, while reptiles fared better, being ecologically adapted to the drier conditions that followed.

An array of Neolithic artifacts, including bracelets, axe heads, chisels, and polishing tools.

Rainforests once covered 14% of the earth's land surface; now they cover a mere 6% and experts estimate that the last remaining rainforests could be consumed in less than 40 years. Small scale deforestation was practiced by some societies for tens of thousands of years before the beginnings of civilization. The first evidence of deforestation appears in the Mesolithic period. It was probably used to convert closed forests into more open ecosystems favourable to game animals. With the advent of agriculture, larger areas began to be deforested, and fire became the prime tool to clear land for crops. In Europe there is little solid evidence before 7000 BC. Mesolithic foragers used fire to create openings for red deer and wild boar. In Great Britain, shade-tolerant species such as oak and ash are replaced in the pollen record by hazels, brambles, grasses and nettles. Removal of the forests led to decreased transpiration, resulting in the formation of upland peat bogs. Widespread decrease in elm pollen across Europe between 8400–8300 BC and 7200–7000 BC, starting in southern Europe and gradually moving north to Great Britain, may represent land clearing by fire at the onset of Neolithic agriculture.

The Neolithic period saw extensive deforestation for farming land. Stone axes were being made from about 3000 BC not just from flint, but from a wide variety of hard rocks from across Britain and North America as well. They include the noted Langdale axe industry in the English Lake District, quarries developed at Penmaenmawr in North Wales and numerous other locations. Rough-outs were made locally near the quarries, and some were polished locally to give a fine finish. This step not only increased the mechanical strength of the axe, but also made penetration of wood easier. Flint was still used from sources such as Grimes Graves but from many other mines across Europe.

Evidence of deforestation has been found in Minoan Crete; for example the environs of the Palace of Knossos were severely deforested in the Bronze Age.

Pre-industrial History

Throughout most of history, humans were hunter gatherers who hunted within forests. In most areas, such as the Amazon, the tropics, Central America, and the Caribbean, only after shortages of wood and other forest products occur are policies implemented to ensure forest resources are used in a sustainable manner.

In ancient Greece, Tjeered van Andel and co-writers summarized three regional studies of historic erosion and alluviation and found that, wherever adequate evidence exists, a major phase of erosion follows, by about 500-1,000 years the introduction of farming in the various regions of Greece, ranging from the later Neolithic to the Early Bronze Age. The thousand years following the mid-first millennium BC saw serious, intermittent pulses of soil erosion in numerous places. The historic silting of ports along the southern coasts of Asia Minor (*e.g.* Clarus, and the examples of Ephesus, Priene and Miletus, where harbors had to be abandoned because of the silt deposited by the Meander) and in coastal Syria during the last centuries BC.

Easter Island

Easter Island has suffered from heavy soil erosion in recent centuries, aggravated by agriculture and deforestation. Jared Diamond gives an extensive look into the collapse of the ancient Easter Islanders in his book *Collapse*. The disappearance of the island's trees seems to coincide with a decline of its civilization around the 17th and 18th century. He attributed the collapse to deforestation and over-exploitation of all resources.

The famous silting up of the harbor for Bruges, which moved port commerce to Antwerp, also followed a period of increased settlement growth (and apparently of deforestation) in the upper river basins. In early medieval Riez in upper Provence, alluvial silt from two small rivers raised the riverbeds and widened the floodplain, which slowly buried the Roman settlement in alluvium and gradually moved new construction to higher ground; concurrently the headwater valleys above Riez were being opened to pasturage.

A typical progress trap was that cities were often built in a forested area, which would provide wood for some industry (for example, construction, shipbuilding, pottery). When deforestation occurs without proper replanting, however; local wood supplies become difficult to obtain near enough to remain competitive, leading to the city's abandonment, as happened repeatedly in Ancient Asia Minor. Because of fuel needs, mining and metallurgy often led to deforestation and city abandonment.

With most of the population remaining active in (or indirectly dependent on) the agricultural sector, the main pressure in most areas remained land clearing for crop and cattle farming. Enough wild green was usually left standing (and partially used, for example, to collect firewood, timber and fruits, or to graze pigs) for wildlife to remain viable. The elite's (nobility and higher clergy) protection of their own hunting privileges and game often protected significant woodlands.

Major parts in the spread (and thus more durable growth) of the population were played by monastical 'pioneering' (especially by the Benedictine and Commercial orders) and some feudal lords' recruiting farmers to settle (and become tax payers) by offering relatively good legal and fiscal conditions. Even when speculators sought to encourage towns, settlers needed an agricultural belt around or sometimes within defensive walls. When populations were quickly decreased by causes such as the Black Death or devastating warfare (for example, Genghis Khan's Mongol hordes in eastern and central Europe, Thirty Years' War in Germany), this could lead to settlements being abandoned. The land was reclaimed by nature, but the secondary forests usually lacked the original biodiversity.

Deforestation of Brazil's Atlantic Forest c.1820-1825

From 1100 to 1500 AD, significant deforestation took place in Western Europe as a result of the expanding human population. The large-scale building of wooden sailing ships by European (coastal) naval owners since the 15th century for exploration, colonisation, slave trade—and other trade on the high seas consumed many forest resources. Piracy also contributed to the over harvesting of forests, as in Spain. This led to a weakening of the domestic economy after Columbus' discovery of America, as the economy became dependent on colonial activities (plundering, mining, cattle, plantations, trade, etc.)

In *Changes in the Land* (1983), William Cronon analyzed and documented 17th-century English colonists' reports of increased seasonal flooding in New England during the period when new settlers initially cleared the forests for agriculture. They believed flooding was linked to widespread forest clearing upstream.

The massive use of charcoal on an industrial scale in Early Modern Europe was a new type of consumption of western forests; even in Stuart England, the relatively primitive production of charcoal has already reached an impressive level. Stuart England was so widely deforested that it depended on the Baltic trade for ship timbers, and looked to the untapped forests of New England to supply the need. Each of Nelson's Royal Navy war ships at Trafalgar (1805) required 6,000 mature oaks for its construction. In France, Colbert planted oak forests to supply the French navy in the future. When the oak plantations matured in the mid-19th century, the masts were no longer required because shipping had changed.

Norman F. Cantor's summary of the effects of late medieval deforestation applies equally well to Early Modern Europe:

Europeans had lived in the midst of vast forests throughout the earlier medieval centuries. After 1250 they became so skilled at deforestation that by 1500 they were running short of wood for heating and cooking. They were faced with a nutritional decline because of the elimination of the generous supply of wild game that had inhabited the now-disappearing forests, which throughout medieval times had provided the staple of their carnivorous high-protein diet. By 1500 Europe was on the edge of a fuel and nutritional disaster [from] which it was saved in the sixteenth century only by the burning of soft coal and the cultivation of potatoes and maize.

Industrial Era

In the 19th century, introduction of steamboats in the United States was the cause of deforestation of banks of major rivers, such as the Mississippi River, with increased and more severe flooding one of the environmental results. The steamboat crews cut wood every day from the riverbanks to fuel the steam engines. Between St. Louis and the confluence with the Ohio River to the south, the Mississippi became more wide and shallow, and changed its channel laterally. Attempts to improve navigation by the use of snag pullers often resulted in crews' clearing large trees 100 to 200 feet (61 m) back from the banks. Several French colonial towns of the Illinois Country, such as Kaskaskia, Cahokia and St. Philippe, Illinois were flooded and abandoned in the late 19th century, with a loss to the cultural record of their archeology.

The wholescale clearance of woodland to create agricultural land can be seen in many parts of the world, such as the Central forest-grasslands transition and other areas of the Great Plains of the United States. Specific parallels are seen in the 20th-century deforestation occurring in many developing nations.

Rates of Deforestation

Global deforestation sharply accelerated around 1852. It has been estimated that about half of the Earth's mature tropical forests—between 7.5 million and 8 million km² (2.9 million to 3 million sq mi) of the original 15 million to 16 million km² (5.8 million to 6.2 million sq mi) that until 1947 covered the planet—have now been destroyed. Some scientists have predicted that unless significant measures (such as seeking out and protecting old growth forests that have not been disturbed) are taken on a worldwide basis, by 2030 there will only be 10% remaining, with another 10% in a degraded condition. 80% will have been lost, and with them hundreds of thousands of irreplaceable species. Some cartographers have attempted to illustrate the sheer scale of deforestation by country using a cartogram.

Estimates vary widely as to the extent of tropical deforestation. Scientists estimate that one fifth of the world's tropical rainforest was destroyed between 1960 and 1990. They claim that that rainforests 60 years ago covered 14% of the world's land surface, now only cover 5–7%, and that all tropical forests will be gone by the middle of the 21st century.

Slash-and-burn farming in the state of Rondônia, western Brazil

A 2002 analysis of satellite imagery suggested that the rate of deforestation in the humid tropics (approximately 5.8 million hectares per year) was roughly 23% lower than the most commonly quoted rates. Conversely, a newer analysis of satellite images reveals that deforestation of the Amazon rainforest is twice as fast as scientists previously estimated.

Some have argued that deforestation trends may follow a Kuznets curve, which if true would nonetheless fail to eliminate the risk of irreversible loss of non-economic forest values (for example, the extinction of species).

Satellite image of Haiti's border with the Dominican Republic (right) shows the amount of deforestation on the Haitian side

A 2005 report by the United Nations Food and Agriculture Organization (FAO) estimates that although the Earth's total forest area continues to decrease at about 13 million hectares per year, the global rate of deforestation has recently been slowing. Still others claim that rainforests are being destroyed at an ever-quickening pace. The London-based Rainforest Foundation notes that "the UN figure is based on a definition of forest as being an area with as little as 10% actual tree cover, which would therefore include areas that are actually savannah-like ecosystems and badly damaged forests." Other critics of the FAO data point out that they do not distinguish between forest types, and that they are based largely on reporting from forestry departments of individual countries, which do not take into account unofficial activities like illegal logging.

Despite these uncertainties, there is agreement that destruction of rainforests remains a significant environmental problem. Up to 90% of West Africa's coastal rainforests have disappeared since 1900. In South Asia, about 88% of the rainforests have been lost. Much of what remains of the world's rainforests is in the Amazon basin, where the Amazon Rainforest covers approximately 4 million square kilometres. The regions with the highest tropical deforestation rate between 2000 and 2005 were Central America—which lost 1.3% of its forests each year—and tropical Asia. In Central America, two-thirds of lowland tropical forests have been turned into pasture since 1950 and 40% of all the rainforests have been lost in the last 40 years. Brazil has lost 90–95% of its Mata Atlântica forest. Paraguay was losing its natural semi humid forests in the country's western regions at a rate of 15.000 hectares at a randomly studied 2-month period in 2010, Paraguay's parliament refused in 2009 to pass a law that would have stopped cutting of natural forests altogether.

Deforestation around Pakke Tiger Reserve, India

Madagascar has lost 90% of its eastern rainforests. As of 2007, less than 1% of Haiti's forests remained. Mexico, India, the Philippines, Indonesia, Thailand, Burma, Malaysia, Bangladesh, China, Sri Lanka, Laos, Nigeria, the Democratic Republic of the Congo, Liberia, Guinea, Ghana and the Ivory Coast, have lost large areas of their rainforest. Several countries, notably Brazil, have declared their deforestation a national emergency. The World Wildlife Fund's ecoregion project catalogues habitat types throughout the world, including habitat loss such as deforestation, showing for example that even in the rich forests of parts of Canada such as the Mid-Continental Canadian forests of the prairie provinces half of the forest cover has been lost or altered.

Regions

Rates of deforestation vary around the world.

In 2011 Conservation International listed the top 10 most endangered forests, characterized by having all lost 90% or more of their original habitat, and each harboring at least 1500 endemic plant species (species found nowhere else in the world).

Top 10 Most Endangered Forests 2011				
Endangered forest	Region	Remaining habitat	Predominate vegetation type	Notes
Indo-Burma	Asia-Pacific	5%	Tropical and subtropical moist broadleaf forests	Rivers, floodplain wetlands, mangrove forests. Burma, Thailand, Laos, Vietnam, Cambodia, India.
New Caledonia	Asia-Pacific	5%	Tropical and subtropical moist broadleaf forests	See note for region covered.
Sundaland	Asia-Pacific	7%	Tropical and subtropical moist broadleaf forests	Western half of the Indo-Malayan archipelago including southern Borneo and Sumatra.
Philippines	Asia-Pacific	7%	Tropical and subtropical moist broadleaf forests	Forests over the entire country including 7,100 islands.
Atlantic Forest	South America	8%	Tropical and subtropical moist broadleaf forests	Forests along Brazil's Atlantic coast, extends to parts of Paraguay, Argentina and Uruguay.
Mountains of Southwest China	Asia-Pacific	8%	Temperate coniferous forest	See note for region covered.
California Floristic Province	North America	10%	Tropical and subtropical dry broadleaf forests	See note for region covered.
Coastal Forests of Eastern Africa	Africa	10%	Tropical and subtropical moist broadleaf forests	Mozambique, Tanzania, Kenya, Somalia.
Madagascar & Indian Ocean Islands	Africa	10%	Tropical and subtropical moist broadleaf forests	Madagascar, Mauritius, Reunion, Seychelles, Comoros.
Eastern Afromontane	Africa	11%	Tropical and subtropical moist broadleaf forests Montane grasslands and shrublands	Forests scattered along the eastern edge of Africa, from Saudi Arabia in the north to Zimbabwe in the south.

Control

Reducing Emissions

Main international organizations including the United Nations and the World Bank, have begun to develop programs aimed at curbing deforestation. The blanket term Reducing Emissions from Deforestation and Forest Degradation (REDD) describes these sorts of programs, which use direct monetary or other incentives to encourage developing countries to limit and/or roll back deforestation. Funding has been an issue, but at the UN Framework Convention on Climate Change (UNFCCC) Conference of the Parties-15 (COP-15) in Copenhagen in December 2009, an accord was reached with a collective commitment by developed countries for new and additional resources, including forestry and investments through international institutions, that will approach USD 30 billion for the period 2010–2012. Significant work is underway on tools for use in monitoring developing country adherence to their agreed REDD targets. These tools, which rely on remote forest monitoring using satellite imagery and other data sources, include the Center for Global Development's FORMA (Forest Monitoring for Action) initiative and the Group on Earth Observations' Forest Carbon Tracking Portal. Methodological guidance for forest monitoring was also emphasized at COP-15. The environmental organization Avoided Deforestation Partners leads the campaign for development of REDD through funding from the U.S. government. In 2014, the Food and Agriculture Organization of the United Nations and partners launched Open Foris - a set of open-source software tools that assist countries in gathering, producing and disseminating information on the state of forest resources. The tools support the inventory lifecycle, from needs assessment, design, planning, field data collection and management, estimation analysis, and dissemination. Remote sensing image processing tools are included, as well as tools for international reporting for Reducing emissions from deforestation and forest degradation (REDD) and MRV and FAO's Global Forest Resource Assessments.

In evaluating implications of overall emissions reductions, countries of greatest concern are those categorized as High Forest Cover with High Rates of Deforestation (HFHD) and Low Forest Cover with High Rates of Deforestation (LFHD). Afghanistan, Benin, Botswana, Burma, Burundi, Cameroon, Chad, Ecuador, El Salvador, Ethiopia, Ghana, Guatemala, Guinea, Haiti, Honduras, Indonesia, Liberia, Malawi, Mali, Mauritania, Mongolia, Namibia, Nepal, Nicaragua, Niger, Nigeria, Pakistan, Paraguay, Philippines, Senegal, Sierra Leone, Sri Lanka, Sudan, Togo, Uganda, United Republic of Tanzania, Zimbabwe are listed as having Low Forest Cover with High Rates of Deforestation (LFHD). Brazil, Cambodia, Democratic Peoples Republic of Korea, Equatorial Guinea, Malaysia, Solomon Islands, Timor-Leste, Venezuela, Zambia are listed as High Forest Cover with High Rates of Deforestation (HFHD).

Payments for Conserving Forests

In Bolivia, deforestation in upper river basins has caused environmental problems, including soil erosion and declining water quality. An innovative project to try and remedy this situation involves landholders in upstream areas being paid by downstream water users to conserve forests. The landholders receive US$20 to conserve the trees, avoid polluting livestock practices, and enhance the biodiversity and forest carbon on their land. They also receive US$30, which purchases a beehive, to compensate for conservation for two hectares of water-sustaining forest for five years. Honey revenue per hectare of forest is US$5 per year, so within five years, the landholder has sold US$50 of honey. The project is being conducted by Fundación Natura Bolivia and Rare Conservation, with support from the Climate & Development Knowledge Network.

Farming

New methods are being developed to farm more intensively, such as high-yield hybrid crops, greenhouse, autonomous building gardens, and hydroponics. These methods are often dependent on chemical inputs to maintain necessary yields. In cyclic agriculture, cattle are grazed on farm land that is resting and rejuvenating. Cyclic agriculture actually increases the fertility of the soil. Intensive farming can also decrease soil nutrients by consuming at an accelerated rate the trace minerals needed for crop growth. The most promising approach, however, is the concept of food forests in permaculture, which consists of agroforestal systems carefully designed to mimic natural forests, with an emphasis on plant and animal species of interest for food, timber and other uses. These systems have low dependence on fossil fuels and agro-chemicals, are highly self-maintaining, highly productive, and with strong positive impact on soil and water quality, and biodiversity.

Monitoring Deforestation

There are multiple methods that are appropriate and reliable for reducing and monitoring deforestation. One method is the "visual interpretation of aerial photos or satellite imagery that is labor-intensive but does not require high-level training in computer image processing or extensive computational resources". Another method includes hot-spot analysis (that is, locations of rapid change) using expert opinion or coarse resolution satellite data to identify locations for detailed digital analysis with high resolution satellite images. Deforestation is typically assessed by quantifying the amount of area deforested, measured at the present time. From an environmental point of view, quantifying the damage and its possible consequences is a more important task, while conservation efforts are more focused on forested land protection and development of land-use alternatives to avoid continued deforestation. Deforestation rate and total area deforested, have been widely used for monitoring deforestation in many regions, including the Brazilian Amazon deforestation monitoring by INPE.

Forest Management

Efforts to stop or slow deforestation have been attempted for many centuries because it has long been known that deforestation can cause environmental damage sufficient in some cases to cause societies to collapse. In Tonga, paramount rulers developed policies designed to prevent conflicts between short-term gains from converting forest to farmland and long-term problems forest loss would cause, while during the 17th and 18th centuries in Tokugawa, Japan, the shoguns developed a highly sophisticated system of long-term planning to stop and even reverse deforestation of the preceding centuries through substituting timber by other products and more efficient use of land that had been farmed for many centuries. In 16th-century Germany, landowners also developed silviculture to deal with the problem of deforestation. However, these policies tend to be limited to environments with *good rainfall*, *no dry season* and *very young soils* (through volcanism or glaciation). This is because on older and less fertile soils trees grow too slowly for silviculture to be economic, whilst in areas with a strong dry season there is always a risk of forest fires destroying a tree crop before it matures.

In the areas where "slash-and-burn" is practiced, switching to "slash-and-char" would prevent the rapid deforestation and subsequent degradation of soils. The biochar thus created, given back to the soil, is not only a durable carbon sequestration method, but it also is an extremely beneficial amendment to the soil. Mixed with biomass it brings the creation of terra preta, one of the richest soils on the planet and the only one known to regenerate itself.

Sustainable Practices

Certification, as provided by global certification systems such as Programme for the Endorsement of Forest Certification and Forest Stewardship Council, contributes to tackling deforestation by creating market demand for timber from sustainably managed forests. According to the United Nations Food and Agriculture Organization (FAO), "A major condition for the adoption of sustainable forest management is a demand for products that are produced sustainably and consumer willingness to pay for the higher costs entailed. Certification represents a shift from regulatory approaches to market incentives to promote sustainable forest management. By promoting the positive attributes of forest products from sustainably managed forests, certification focuses on the demand side of environmental conservation." Rainforest Rescue argues that the standards of organizations like FSC are too closely connected to timber industry interests and therefore do not guarantee environmentally and socially responsible forest management. In reality, monitoring systems are inadequate and various cases of fraud have been documented worldwide.

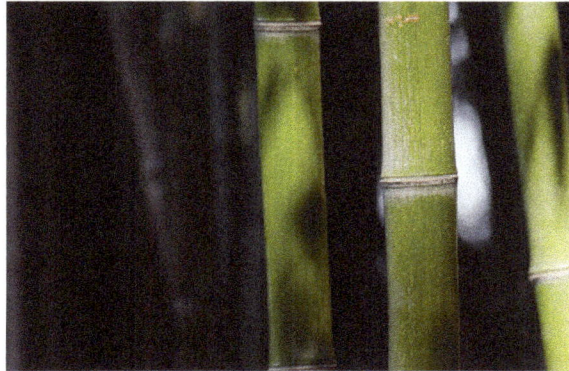

Bamboo is advocated as a more sustainable alternative for cutting down wood for fuel.

Some nations have taken steps to help increase the amount of trees on Earth. In 1981, China created National Tree Planting Day Forest and forest coverage had now reached 16.55% of China's land mass, as against only 12% two decades ago

Using fuel from bamboo rather than wood results in cleaner burning, and since bamboo matures much faster than wood, deforestation is reduced as supply can be replenished faster.

Reforestation

In many parts of the world, especially in East Asian countries, reforestation and afforestation are increasing the area of forested lands. The amount of woodland has increased in 22 of the world's 50 most forested nations. Asia as a whole gained 1 million hectares of forest between 2000 and 2005. Tropical forest in El Salvador expanded more than 20% between 1992 and 2001. Based on these trends, one study projects that global forest will increase by 10%—an area the size of India—by 2050.

In the People's Republic of China, where large scale destruction of forests has occurred, the government has in the past required that every able-bodied citizen between the ages of 11 and 60 plant three to five trees per year or do the equivalent amount of work in other forest services. The government claims that at least 1 billion trees have been planted in China every year since 1982. This is no longer required today, but March 12 of every year in China is the Planting Holiday. Also, it has introduced the Green Wall of China project, which aims to halt the expansion of the Gobi desert through the planting of trees. However, due to the large percentage of trees dying off after planting (up to 75%), the project is not very successful. There has been a 47-million-hectare increase in forest area in China since the 1970s. The total number of trees amounted to be about 35 billion and 4.55% of China's land mass increased in forest coverage. The forest coverage was 12% two decades ago and now is 16.55%.

An ambitious proposal for China is the Aerially Delivered Re-forestation and Erosion Control System and the proposed Sahara Forest Project coupled with the Seawater Greenhouse.

In Western countries, increasing consumer demand for wood products that have been produced and harvested in a sustainable manner is causing forest landowners and forest industries to become increasingly accountable for their forest management and timber harvesting practices.

The Arbor Day Foundation's Rain Forest Rescue program is a charity that helps to prevent deforestation. The charity uses donated money to buy up and preserve rainforest land before the lumber companies can buy it. The Arbor Day Foundation then protects the land from deforestation. This also locks in the way of life of the primitive tribes living on the forest land. Organizations such as Community Forestry International, Cool Earth, The Nature Conservancy, World Wide Fund for Nature, Conservation International, African Conservation Foundation and Greenpeace also focus on preserving forest habitats. Greenpeace in particular has also mapped out the forests that are still intact and published this information on the internet. World Resources Institute in turn has made a simpler thematic map showing the amount of forests present just before the age of man (8000 years ago) and the current (reduced) levels of forest. These maps mark the amount of afforestation required to repair the damage caused by people.

Forest Plantations

To meet the world's demand for wood, it has been suggested by forestry writers Botkins and Sedjo that high-yielding forest plantations are suitable. It has been calculated that plantations yielding 10 cubic meters per hectare annually could supply all the timber required for international trade on 5% of the world's existing forestland. By contrast, natural forests produce about 1–2 cubic meters per hectare; therefore, 5–10 times more forestland would be required to meet demand. Forester Chad Oliver has suggested a forest mosaic with high-yield forest lands interspersed with conservation land.

In the country of Senegal, on the western coast of Africa, a movement headed by youths has helped to plant over 6 million mangrove trees. The trees will protect local villages from storm damages and will provide a habitat for local wildlife. The project started in 2008, and already the Senegalese government has been asked to establish rules and regulations that would protect the new mangrove forests.

Military Context

While the preponderance of deforestation is due to demands for agricultural and urban use for the human population, there are some examples of military causes. One example of deliberate deforestation is that which took place in the U.S. zone of occupation in Germany after World War II. Before the onset of the Cold War, defeated Germany was still considered a potential future threat rather than potential future ally. To address this threat, attempts were made to lower German industrial potential, of which forests were deemed an element. Sources in the U.S. government admitted that the purpose of this was that the "ultimate destruction of the war potential of German forests." As a consequence of the practice of clear-felling, deforestation resulted which could "be replaced only by long forestry development over perhaps a century."

American Sherman tanks knocked out by Japanese artillery on Okinawa.

Deforestation can also be one consequence of war. For example, in the 1945 Battle of Okinawa, bombardment and other combat operations reduced the lush tropical landscape into "a vast field of mud, lead, decay and maggots". Deforestation can also be an intentional tactic of military forces. Defoliants (like Agent Orange or others) was used by the British in the Malayan Emergency, and by the United States in the Korean War and Vietnam War.

Public Health Context

Deforestation eliminates a great number of species of plants and animals which also often results in an increase in disease. Loss of native species allows new species to come to dominance. Often the destruction of predatory species can result in an increase in rodent populations. These are known to carry plagues. Additionally, erosion can produce pools of stagnant water that are perfect breeding grounds for mosquitos, well known vectors of malaria, yellow fever, nipah virus, and more. Deforestation can also create a path for non-native species to flourish such as certain types of snails, which have been correlated with an increase in schistosomiasis cases.

Deforestation is occurring all over the world and has been coupled with an increase in the occurrence of disease outbreaks. In Malaysia, thousands of acres of forest have been cleared for pig farms. This has resulted in an increase in the zoonosis the Nipah virus. In Kenya, deforestation has led to an increase in malaria cases which is now the leading cause of morbidity and mortality the country.

Another pathway through which deforestation affects disease is the relocation and dispersion of disease-carrying hosts. This disease emergence pathway can be called "range expansion," whereby the host's range (and thereby the range of pathogens) expands to new geographic areas. Through deforestation, hosts and reservoir species are forced into neighboring habitats. Accompanying the reservoir species are pathogens that have the ability to find new hosts in previously unexposed regions. As these pathogens and species come into closer contact with humans, they are infected both directly and indirectly.

A catastrophic example of range expansion is the 1998 outbreak of Nipah Virus in Malaysia. For a number of years, deforestation, drought, and subsequent fires led to a dramatic geographic shift and density of fruit bats, a reservoir for Nipah virus. Deforestation reduced the available fruiting trees in the bats' habitat, and they encroached on surrounding orchards which also happened to be the location of a large number of pigsties. The bats, through proximity spread the Nipah to pigs. While the virus infected the pigs, mortality was much lower than among humans, making the pigs a virulent host leading to the transmission of the virus to humans. This resulted in 265 reported cases of encephalitis, of which 105 resulted in death. This example provides an important lesson for the impact deforestation can have on human health.

Another example of range expansion due to deforestation and other anthropogenic habitat impacts includes the Capybara rodent in Paraguay. This rodent is the host of a number of zoonotic diseases and, while there has not yet been a human-borne outbreak due to the movement of this rodent into new regions, it offers an example of how habitat destruction through deforestation and subsequent movements of species is occurring regularly.

A now well-developed theory is that the spread of HIV it is at least partially due deforestation. Rising populations created a food demand and with deforestation opening up new areas of the forest the hunters harvested a great deal of primate bushmeat, which is believed to be the origin of HIV.

Habitat Destruction

Habitat destruction is the process in which natural habitat is rendered unable to support the species present. In this process, the organisms that previously used the site are displaced or destroyed, reducing biodiversity. Habitat destruction by human activity is mainly for the purpose of harvesting natural resources for industry production and urbanization. Clearing habitats for agriculture is the principal cause of habitat destruction. Other important causes of habitat destruction include mining, logging, trawling and urban sprawl. Habitat destruction is currently ranked as the primary cause of species extinction worldwide. It is a process of natural environmental change that may be caused by habitat fragmentation, geological processes, climate change or by human activities such as the introduction of invasive species, ecosystem nutrient depletion, and other human activities

The terms habitat loss and habitat reduction are also used in a wider sense, including loss of habitat from other factors, such as water and noise pollution.

Impacts on Organisms

In the simplest term, when a habitat is destroyed, the plants, animals, and other organisms that occupied the habitat have a reduced carrying capacity so that populations decline and extinction becomes more likely. Perhaps the greatest threat to organisms and biodiversity is the process of habitat loss. Temple (1986) found that 82% of endangered bird species were significantly threatened by habitat loss. Endemic organisms with limited ranges are most affected by habitat destruction, mainly because these organisms are not found anywhere else within the world and thus, have less chance of recovering. Many endemic organisms have very specific requirements for their survival that can only be found within a certain ecosystem, resulting in their extinction. Extinction may take place very long after the destruction of habitat, however, a phenomenon known as extinction debt. Habitat destruction can also decrease the range of certain organism populations. This can result in the reduction of genetic diversity and perhaps the production of infertile youths, as these organisms would have a higher possibility of mating with related organisms within their population, or different species. One of the most famous examples is the impact upon China's giant panda, once found across the nation. Now it is only found in fragmented and isolated regions in the south-west of the country, as a result of widespread deforestation in the 20th Century.

Geography

Biodiversity hotspots are chiefly tropical regions that feature high concentrations of endemic species and, when all hotspots are combined, may contain over half of the world's terrestrial species. These hotspots are suffering from habitat loss and destruction. Most of the natural habitat on islands and in areas of high human population density has already been destroyed (WRI, 2003). Islands suffering extreme habitat destruction include New Zealand, Madagascar, the Philippines, and Japan. South and east Asia — especially China, India, Malaysia, Indonesia, and Japan — and many areas in West Africa have extremely dense human populations that allow little room for natural habitat. Marine areas close to highly populated coastal cities also face degradation of their coral reefs or other marine habitat. These areas include the eastern coasts of Asia and Africa, northern coasts of South America, and the Caribbean Sea and its associated islands.

Satellite photograph of deforestation in Bolivia. Originally dry tropical forest, the land is being cleared for soybean cultivation.

Regions of unsustainable agriculture or unstable governments, which may go hand-in-hand, typically experience high rates of habitat destruction. Central America, Sub-Saharan Africa, and the Amazonian tropical rainforest areas of South America are the main regions with unsustainable agricultural practices or government mismanagement.

Areas of high agricultural output tend to have the highest extent of habitat destruction. In the U.S., less than 25% of native vegetation remains in many parts of the East and Midwest. Only 15% of land area remains unmodified by human activities in all of Europe.

Ecosystems

Tropical rainforests have received most of the attention concerning the destruction of habitat. From the approximately 16 million square kilometers of tropical rainforest habitat that originally existed worldwide, less than 9 million square kilometers remain today. The current rate of deforestation is 160,000 square kilometers per year, which equates to a loss of approximately 1% of original forest habitat each year.

Jungle burned for agriculture in southern Mexico

Other forest ecosystems have suffered as much or more destruction as tropical rainforests. Farming and logging have severely disturbed at least 94% of temperate broadleaf forests; many old growth forest stands have lost more than 98% of their previous area because of human activities. Tropical deciduous dry forests are easier to clear and burn and are more suitable for agriculture and cattle ranching than tropical rainforests; consequently, less than 0.1% of dry forests in Central America's Pacific Coast and less than 8% in Madagascar remain from their original extents.

Farmers near newly cleared land within Taman Nasional Kerinci Seblat (Kerinci Seblat National Park), Sumatra.

Plains and desert areas have been degraded to a lesser extent. Only 10-20% of the world's drylands, which include temperate grasslands, savannas, and shrublands, scrub, and deciduous forests, have been somewhat degraded. But included in that 10-20% of land is the approximately 9 million square kilometers of seasonally dry-lands that humans have converted to deserts through the process of desertification. The tallgrass prairies of North America, on the other hand, have less than 3% of natural habitat remaining that has not been converted to farmland.

Wetlands and marine areas have endured high levels of habitat destruction. More than 50% of wetlands in the U.S. have been destroyed in just the last 200 years. Between 60% and 70% of European wetlands have been completely destroyed. About one-fifth (20%) of marine coastal areas have been highly modified by humans. One-fifth of coral reefs have also been destroyed, and another fifth has been severely degraded by overfishing, pollution, and invasive species; 90% of the Philippines' coral reefs alone have been destroyed. Finally, over 35% mangrove ecosystems worldwide have been destroyed.

Natural Causes

Habitat destruction through natural processes such as volcanism, fire, and climate change is well documented in the fossil record. One study shows that habitat fragmentation of tropical rainforests in Euramerica 300 million years ago led to a great loss of amphibian diversity, but simultaneously the drier climate spurred on a burst of diversity among reptiles.

Human Causes

Habitat destruction caused by humans includes land conversion from forests, etc. to arable land, urban sprawl, infrastructure development, and other anthropogenic changes to the characteristics of land. Habitat degradation, fragmentation, and pollution are aspects of habitat destruction caused by humans that do not necessarily involve over destruction of habitat, yet result in habitat collapse. Desertification, deforestation, and coral reef degradation are specific types of habitat destruction for those areas (deserts, forests, coral reefs).

Geist and Lambin (2002) assessed 152 case studies of net losses of tropical forest cover to determine any patterns in the proximate and underlying causes of tropical deforestation. Their results, yielded as percentages of the case studies in which each parameter was a significant factor, provide a quantitative prioritization of which proximate and underlying causes were the most significant. The proximate causes were clustered into broad categories of agricultural expansion (96%), infrastructure expansion (72%), and wood extraction (67%). Therefore, according to this study, forest conversion to agriculture is the main land use change responsible for tropical deforestation. The specific categories reveal further insight into the specific causes of tropical deforestation: transport extension (64%), commercial wood extraction (52%), permanent cultivation (48%), cattle ranching (46%), shifting (slash and burn) cultivation (41%), subsistence agriculture (40%), and fuel wood extraction for domestic use (28%). One result is that shifting cultivation is not the primary cause of deforestation in all world regions, while transport extension (including the construction of new roads) is the largest single proximate factor responsible for deforestation.

Drivers

While the above-mentioned activities are the proximal or direct causes of habitat destruction in that they actually destroy habitat, this still does not identify why humans destroy habitat. The forces that cause humans to destroy habitat are known as *drivers* of habitat destruction. Demographic, economic, sociopolitical, scientific and technological, and cultural drivers all contribute to habitat destruction.

Demographic drivers include the expanding human population; rate of population increase over time; spatial distribution of people in a given area (urban versus rural), ecosystem type, and country; and the combined effects of poverty, age, family planning, gender, and education status of people in certain areas. Most of the exponential human population growth worldwide is occurring in or close to biodiversity hotspots. This may explain why human population density accounts for 87.9% of the variation in numbers of threatened species across 114 countries, providing indisputable evidence that people play the largest role in decreasing biodiversity. The boom in human population and migration of people into such species-rich regions are making conservation efforts not only more urgent but also more likely to conflict with local human interests. The high local population density in such areas is directly correlated to the poverty status of the local people, most of whom lacking an education and family planning.

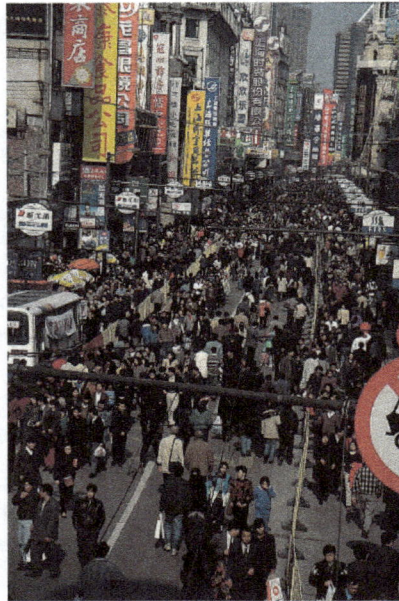

Nanjing Road in Shanghai

From the Geist and Lambin (2002) study described in the previous section, the underlying driving forces were prioritized as follows (with the percent of the 152 cases the factor played a significant role in): economic factors (81%), institutional or policy factors (78%), technological factors (70%), cultural or socio-political factors (66%), and demographic factors (61%). The main economic factors included commercialization and growth of timber markets (68%), which are driven by national and international demands; urban industrial growth (38%); low domestic costs for land, labor, fuel, and timber (32%); and increases in product prices mainly for cash crops (25%). Institutional and policy factors included formal pro-deforestation policies on land development (40%), economic growth including colonization and infrastructure improvement (34%), and subsidies for land-based activities (26%); property rights and land-tenure insecurity (44%); and policy failures such as corruption, lawlessness, or mismanagement (42%). The main technological factor was the poor application of technology in the wood industry (45%), which leads to wasteful logging practices. Within the broad category of cultural and sociopolitical factors are public attitudes and values (63%), individual/household behavior (53%), public unconcern toward forest environments (43%), missing basic values (36%), and unconcern by individuals (32%). Demographic factors were the in-migration of colonizing settlers into sparsely populated forest areas (38%) and growing population density — a result of the first factor — in those areas (25%).

There are also feedbacks and interactions among the proximate and underlying causes of deforestation that can amplify the process. Road construction has the largest feedback effect, because it interacts with—and leads to—the establishment of new settlements and more people, which causes a growth in wood (logging) and food markets. Growth in these markets, in turn, progresses the commercialization of agriculture and logging industries. When these industries become commercialized, they must become more efficient by utilizing larger or more modern machinery that often are worse on the habitat than traditional farming and logging methods. Either way, more land is cleared more rapidly for commercial markets. This common feedback example manifests just how closely related the proximate and underlying causes are to each other.

Impact on Human Population

Habitat destruction vastly increases an area's vulnerability to natural disasters like flood and drought, crop failure, spread of disease, and water contamination. On the other hand, a healthy ecosystem with good management practices will reduce the chance of these events happening, or will at least mitigate adverse impacts.

The draining and development of coastal wetlands that previously protected the Gulf Coast contributed to severe flooding in New Orleans, Louisiana in the aftermath of Hurricane Katrina.

Agricultural land can actually suffer from the destruction of the surrounding landscape. Over the past 50 years, the destruction of habitat surrounding agricultural land has degraded approximately 40% of agricultural land worldwide via erosion, salinization, compaction, nutrient depletion, pollution, and urbanization. Humans also lose direct uses of natural habitat when habitat is destroyed. Aesthetic uses such as birdwatching, recreational uses like hunting and fishing, and ecotourism usually rely upon virtually undisturbed habitat. Many people value the complexity of the natural world and are disturbed by the loss of natural habitats and animal or plant species worldwide.

Probably the most profound impact that habitat destruction has on people is the loss of many valuable ecosystem services. Habitat destruction has altered nitrogen, phosphorus, sulfur, and carbon cycles, which has increased the frequency and severity of acid rain, algal blooms, and fish kills in rivers and oceans and contributed tremendously to global climate change. One ecosystem service whose significance is becoming more realized is climate regulation. On a local scale, trees provide windbreaks and shade; on a regional scale, plant transpiration recycles rainwater and maintains constant annual rainfall; on a global scale, plants (especially trees from tropical rainforests) from around the world counter the accumulation of greenhouse gases in the atmosphere by sequestering carbon dioxide through photosynthesis. Other ecosystem services that are diminished or lost alto-gether as a result of habitat destruction include watershed management, nitrogen fixation, oxygen production, pollination, waste treatment (i.e., the breaking down and immobilization of toxic pol-lutants), and nutrient recycling of sewage or agricultural runoff.

The loss of trees from the tropical rainforests alone represents a substantial diminishing of the earth's ability to produce oxygen and use up carbon dioxide. These services are becoming even more important as increasing carbon dioxide levels is one of the main contributors to global cli-mate change.

The loss of biodiversity may not directly affect humans, but the indirect effects of losing many species as well as the diversity of ecosystems in general are enormous. When biodiversity is lost, the environment loses many species that provide valuable and unique roles to the ecosystem. The environment and all its inhabitants rely on biodiversity to recover from extreme environmental conditions. When too much biodiversity is lost, a catastrophic event such as an earthquake, flood, or volcanic eruption could cause an ecosystem to crash, and humans would obviously suffer from that. Loss of biodiversity also means that humans are losing animals that could have served as biological control agents and plants that could potentially provide higher-yielding crop varieties, pharmaceutical drugs to cure existing or future diseases or cancer, and new resistant crop varieties for agricultural species susceptible to pesticide-resistant insects or virulent strains of fungi, virus-es, and bacteria.

The negative effects of habitat destruction usually impact rural populations more directly than ur-ban populations. Across the globe, poor people suffer the most when natural habitat is destroyed, because less natural habitat means less natural resources per capita, yet wealthier people and countries simply have to pay more to continue to receive more than their per capita share of nat-ural resources.

Another way to view the negative effects of habitat destruction is to look at the opportunity cost of keeping an area undisturbed. In other words, what are people losing out on by taking away a given habitat? A country may increase its food supply by converting forest land to row-crop agriculture, but the value of the same land may be much larger when it can supply natural resources or services such as clean water, timber, ecotourism, or flood regulation and drought control.

Outlook

The rapid expansion of the global human population is increasing the world's food requirement substantially. Simple logic instructs that more people will require more food. In fact, as the world's population increases dramatically, agricultural output will need to increase by at least 50%, over the next 30 years. In the past, continually moving to new land and soils provided a boost in food production to appease the global food demand. That easy fix will no longer be available, however, as more than 98% of all land suitable for agriculture is already in use or degraded beyond repair.

The impending global food crisis will be a major source of habitat destruction. Commercial farmers are going to become desperate to produce more food from the same amount of land, so they will use more fertilizers and less concern for the environment to meet the market de-mand. Others will seek out new land or will convert other land-uses to agriculture. Agricultural intensification will become widespread at the cost of the environment and its inhabitants. Species will be pushed out of their habitat either directly by habitat destruction or indirectly by fragmentation, degradation, or pollution. Any efforts to protect the world's remaining natural habitat and biodiversity will compete directly with humans' growing demand for natural resources, especially new agricultural lands.

Solutions

Chelonia mydas on a Hawaiian coral reef. Although the endangered species is protected, habitat loss from human development is a major reason for the loss of green turtle nesting beaches.

In most cases of tropical deforestation, three to four underlying causes are driving two to three proximate causes. This means that a universal policy for controlling tropical deforestation would not be able to address the unique combination of proximate and underlying causes of deforestation in each country. Before any local, national, or international deforestation policies are written and enforced, governmental leaders must acquire a detailed understanding of the complex combination of proximate causes and underlying driving forces of deforestation in a given area or country. This concept, along with many other results about tropical deforestation from the Geist and Lambin study, can easily be applied to habitat destruction in general. Governmental leaders need to take action by addressing the underlying driving forces, rather than merely regulating the proximate causes. In a broader sense, governmental bodies at a local, national, and international scale need to emphasize the following:

1. Considering the many irreplaceable ecosystem services provided by natural habitats.

2. Protecting remaining intact sections of natural habitat.

3. Educating the public about the importance of natural habitat and biodiversity.

4. Developing family planning programs in areas of rapid population growth.

5. Finding ecological ways to increase agricultural output without increasing the total land in production.

6. Preserving habitat corridors to minimize prior damage from fragmented habitats.

7. Reduce human population and expansion.

Genetic Pollution

Genetic pollution is a controversial term for uncontrolled gene flow into wild populations. This gene flow is undesirable according to some environmentalists and conservationists, including groups such as Greenpeace, TRAFFIC, and GeneWatch UK.

Usage

- Some conservation biologists and conservationists have used genetic pollution for a number of years as a term to describe gene flow (which they regard as undesirable) from a domestic, feral, non-native or invasive subspecies to a wild indigenous population.

- The term is of late being associated with the gene flow from a genetically engineered (GE) organism to a non GE organism, frequently by those who consider such gene flow detrimental.

Invasive Species

Conservation biologists and conservationists have, for a number of years, used the term to describe gene flow from domestic, feral, and non-native species into wild indigenous species, which they consider undesirable. For example, TRAFFIC is the international wildlife trade monitoring network that works to limit trade in wild plants and animals so that it is not a threat to conservationist goals. They promote awareness of the effects of introduced invasive species that may "*hybridize with native species, causing genetic pollution*". The Joint Nature Conservation Committee (JNCC) is the statutory adviser to the Government of United Kingdom and international nature conservation. Its work contributes to maintaining and enriching biological diversity and educating about the effects of the introduction of invasive/non-native species. In this context they have advised that invasive species:

> "*will alter the genetic pool (a process called genetic pollution), which is an irreversible change.*"

A classic example of an introduced species creating issues revolving around genetic pollution is the Mallard (*A. Platyrhyncos*), which is able to breed with other duck species and create fertile hybrids, introducing unwanted genes into the populations of other wild ducks.

Genetic Engineering

In the fields of agriculture, agroforestry and animal husbandry, *genetic pollution* is being used to describe gene flows between GE species and wild relatives. An early use of the term *genetic pollution* in this later sense appears in a wide-ranging review of the potential ecological effects of genetic engineering in The Ecologist magazine in July 1989. It was also popularized by environmentalist Jeremy Rifkin in his 1998 book *The Biotech Century*. While intentional crossbreeding between two genetically distinct varieties is described as hybridization with the subsequent introgression of genes, Rifkin, who had played a leading role in the ethical debate for over a decade before, used genetic pollution to describe what he considered to be problems that might occur due the unintentional process of (modernly) genetically modified organisms (GMOs) dispersing their genes into the natural environment by breeding with wild plants or animals.

The usage of genetic pollution by the Food and Agriculture Organization of the United Nations (FAO) is currently defined as:

> *"Uncontrolled spread of genetic information (frequently referring to transgenes) into the genomes of organisms in which such genes are not present in nature."*

Since 2005 there has existed a GM Contamination Register, launched for GeneWatch UK and Greenpeace International that records all incidents of intentional or accidental release of organisms genetically modified using modern techniques.

In a 10-year study of four different crops, none of the genetically engineered plants were found to be more invasive or more persistent than their conventional counterparts. An often cited claimed example of genetic pollution is the reputed discovery of transgenes from GE maize in landraces of maize in Oaxaca, Mexico. The report from Quist and Chapela, has since been discredited on methodological grounds. The scientific journal that originally published the study concluded that "the evidence available is not sufficient to justify the publication of the original paper." More recent attempts to replicate the original studies have concluded that genetically modified corn is absent from southern Mexico in 2003 and 2004.

A 2009 study verified the original findings of the controversial 2001 study, by finding transgenes in about 1% of 2000 samples of wild maize in Oaxaca, Mexico, despite Nature retracting the 2001 study and a second study failing to back up the findings of the initial study. The study found that the transgenes are common in some fields, but non-existent in others, hence explaining why a previous study failed to find them. Furthermore, not every laboratory method managed to find the transgenes.

A 2004 study performed near an Oregon field trial for a genetically modified variety of creeping bentgrass (*Agrostis stolonifera*) revealed that the transgene and its associate trait (resistance to the glyphosate herbicide) could be transmitted by wind pollination to resident plants of different *Agrostis* species, up to 14 km from the test field. In 2007, the Scotts Company, producer of the genetically modified bentgrass, agreed to pay a civil penalty of $500,000 to the United States Department of Agriculture (USDA). The USDA alleged that Scotts "failed to conduct a 2003 Oregon field trial in a manner which ensured that neither glyphosate-tolerant creeping bentgrass nor its offspring would persist in the environment".

Controversial Term

Whether genetic pollution or similar terms, such as *"genetic deterioration"*, *"genetic swamping"*, *"genetic takeover"* and *"genetic aggression"*, are an appropriate scientific description of the biology of invasive species is debated. Rhymer and Simberloff argue that these types of terms:

> ...imply either that hybrids are less fit than the parentals, which need not be the case, or that there is an inherent value in "pure" gene pools.

They recommend that gene flow from invasive species be termed genetic mixing since:

> "Mixing" need not be value-laden, and we use it here to denote mixing of gene pools whether or not associated with a decline in fitness.

Environmentalists such as Patrick Moore, an ex-member and cofounder of Greenpeace, questions if the term genetic pollution is more political than scientific. The term is considered to arouse emotional feelings towards the subject matter. In an interview he comments:

> If you take a term used quite frequently these days, the term "genetic pollution," otherwise referred to as genetic contamination, it is a propaganda term, not a technical or scientific term. Pollution and contamination are both value judgments. By using the word "genetic" it gives the public the impression that they are talking about something scientific or technical--as if there were such a thing as genes that amount to pollution.

Overexploitation

Overexploitation, also called overharvesting, refers to harvesting a renewable resource to the point of diminishing returns. Sustained overexploitation can lead to the destruction of the resource. The term applies to natural resources such as: wild medicinal plants, grazing pastures, game animals, fish stocks, forests, and water aquifers.

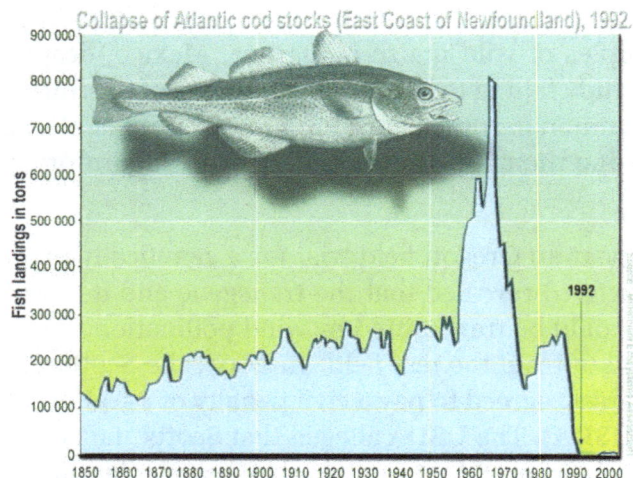

Atlantic cod stocks were severely overexploited in the 1970s and 1980s, leading to their abrupt collapse in 1992.

In ecology, overexploitation describes one of the five main activities threatening global biodiversity. Ecologists use the term to describe populations that are harvested at a rate that is unsustainable, given their natural rates of mortality and capacities for reproduction. This can result in extinction at the population level and even extinction of whole species. In conservation biology the term is usually used in the context of human economic activity that involves the taking of biological resources, or organisms, in larger numbers than their populations can withstand. The term is also used and defined somewhat differently in fisheries, hydrology and natural resource management.

Overexploitation can lead to resource destruction, including extinctions. However it is also possible for overexploitation to be sustainable, as discussed below in the section on fisheries. In the context of fishing, the term overfishing can be used instead of overexploitation, as can overgrazing in stock management, overlogging in forest management, overdrafting in aquifer management, and endangered species in species monitoring. Overexploitation is not an activity limited to humans. Introduced predators and herbivores, for example, can overexploit native flora and fauna.

History

Concern about overexploitation is relatively recent, though overexploitation itself is not a new phenomenon. It has been observed for millennia. For example, ceremonial cloaks worn by the Hawaiian kings were made from the mamo bird; a single cloak used the feathers of 70,000 birds of this now-extinct species. The dodo, a flightless bird from Mauritius, is another well-known example of overexploitation. As with many island species, it was naive about certain predators, allowing humans to approach and kill it with ease.

When the giant flightless birds called moa were overexploited to the point of extinction, the giant Haast's eagle that preyed on them also became extinct

From the earliest of times, hunting has been an important human activity as a means of survival. There is a whole history of overexploitation in the form of overhunting. The overkill hypothesis (Quaternary extinction events) explains why the megafaunal extinctions occurred within a relatively short period of time. This can be traced with human migration. The most convincing evidence of this theory is that 80% of the North American large mammal species disappeared within 1000 years of the arrival of humans on the western hemisphere continents. The fastest ever recorded extinction of megafauna occurred in New Zealand, where by 1500 AD, just 200 years after settling the islands, ten species of the giant moa birds were hunted to extinction by the Māori. A second wave of extinctions occurred later with European settlement.

In more recent times, overexploitation has resulted in the gradual emergence of the concepts of sustainability and sustainable development, which has built on other concepts, such as sustainable yield, eco-development and deep ecology.

Overview

Overexploitation doesn't necessarily lead to the destruction of the resource, nor is it necessarily unsustainable. However, depleting the numbers or amount of the resource can change its quality. For example, footstool palm is a wild palm tree found in Southeast Asia. Its leaves are used for thatching and food wrapping, and overharvesting has resulted in its leaf size becoming smaller.

Tragedy of the Commons

The tragedy of the commons refers to a dilemma described in an article by that name written by Garrett Hardin and first published in the journal *Science* in 1968.

Cows on Selsley Common. The tragedy of the commons is a useful parable for understanding how overexploitation can occur

Central to Hardin's essay is an example which is a useful parable for understanding how over-exploitation can occur. This example was first sketched in an 1833 pamphlet by William Forster Lloyd, as a hypothetical and simplified situation based on medieval land tenure in Europe, of herders sharing a common on which they are each entitled to let their cows graze. In Hardin's example, it is in each herder's interest to put each succeeding cow he acquires onto the land, even if the carrying capacity of the common is exceeded and it is temporarily or permanently damaged for all as a result. The herder receives all of the benefits from an additional cow, while the damage to the common is shared by the entire group. If all herders make this individually rational economic decision, the common will be overexploited or even destroyed to the detriment of all. However, since all herders reach the same rational conclusion, overexploitation in the form of overgrazing occurs, with immediate losses, and the pasture may be degraded to the point where it gives very little return.

"Therein is the tragedy. Each man is locked into a system that compels him to increase his herd without limit—in a world that is limited. Ruin is the destination toward which all men rush, each pursuing his own interest in a society that believes in the freedom of the commons." (Hardin, 1968)

In the course of his essay, Hardin develops the theme, drawing in many examples of latter day commons, such as national parks, the atmosphere, oceans, rivers and fish stocks. The example of fish stocks had led some to call this the "tragedy of the fishers". A major theme running through the essay is the growth of human populations, with the Earth's finite resources being the general common.

The tragedy of the commons has intellectual roots tracing back to Aristotle, who noted that "what is common to the greatest number has the least care bestowed upon it", as well as to Hobbes and his *Leviathan*. The opposite situation to a tragedy of the commons is sometimes referred to as a tragedy of the anticommons: a situation in which rational individuals, acting separately, collectively waste a given resource by underutilizing it.

The tragedy of the commons can be avoided if it is appropriately regulated. Hardin's use of "commons" has frequently been misunderstood, leading Hardin to later remark that he should have titled his work "The tragedy of the unregulated commons".

Fisheries

In wild fisheries, overexploitation or overfishing occurs when a fish stock has been fished down "below the size that, on average, would support the long-term maximum sustainable yield of the fishery". However, overexploitation can be sustainable.

The Atlantic bluefin tuna is currently seriously overexploited. Scientists say 7,500 tons annually is the sustainable limit, yet the fishing industry continue to harvest 60,000 tons.

When a fishery starts harvesting fish from a previously unexploited stock, the biomass of the fish stock will decrease, since harvesting means fish are being removed. For sustainability, the rate at which the fish replenish biomass through reproduction must balance the rate at which the fish are being harvested. If the harvest rate is increased, then the stock biomass will further decrease. At a certain point, the maximum harvest yield that can be sustained will be reached, and further attempts to increase the harvest rate will result in the collapse of the fishery. This point is called the maximum sustainable yield, and in practice, usually occurs when the fishery has been fished down to about 30% of the biomass it had before harvesting started.

It is possible to fish the stock down further to, say, 15% of the pre-harvest biomass, and then adjust the harvest rate so the biomass remains at that level. In this case, the fishery is sustainable, but is now overexploited, because the stock has been run down to the point where the sustainable yield is less than it could be.

Fish stocks are said to "collapse" if their biomass declines by more than 95 percent of their maximum historical biomass. Atlantic cod stocks were severely overexploited in the 1970s and 1980s, leading to their abrupt collapse in 1992. Even though fishing has ceased, the cod stocks have failed to recover. The absence of cod as the apex predator in many areas has led to trophic cascades.

About 25% of world fisheries are now overexploited to the point where their current biomass is less than the level that maximizes their sustainable yield. These depleted fisheries can often recover if fishing pressure is reduced until the stock biomass returns to the optimal biomass. At this point, harvesting can be resumed near the maximum sustainable yield.

The tragedy of the commons can be avoided within the context of fisheries if fishing effort and practices are regulated appropriately by fisheries management. One effective approach may be assigning some measure of ownership in the form of individual transferable quotas (ITQs) to fishermen. In 2008, a large scale study of fisheries that used ITQs, and ones that didn't, provided strong evidence that ITQs help prevent collapses and restore fisheries that appear to be in decline.

Water Resources

Water resources, such as lakes and aquifers, are usually renewable resources which naturally recharge (the term fossil water is sometimes used to describe aquifers which don't recharge). Overexploitation occurs if a water resource, such as the Ogallala Aquifer, is mined or extracted at a rate that exceeds the recharge rate, that is, at a rate that exceeds the practical sustained yield. Recharge usually comes from area streams, rivers and lakes. An aquifer which has been overexploited is said to be overdrafted or depleted. Forests enhance the recharge of aquifers in some locales, although generally forests are a major source of aquifer depletion. Depleted aquifers can become polluted with contaminants such as nitrates, or permanently damaged through subsidence or through saline intrusion from the ocean.

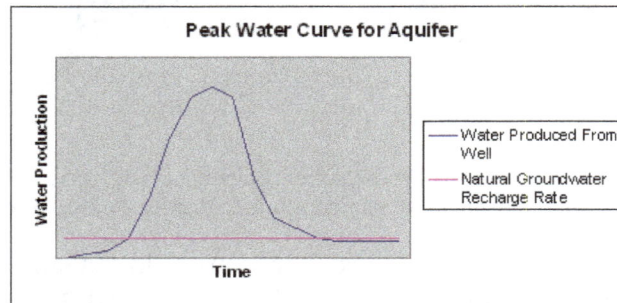

Overexploitation of groundwater from an aquifer can result in a peak water curve.

This turns much of the world's underground water and lakes into finite resources with peak usage debates similar to oil. These debates usually centre around agriculture and suburban water usage but generation of electricity from nuclear energy or coal and tar sands mining is also water resource intensive. A modified Hubbert curve applies to any resource that can be harvested faster than it can be replaced. Though Hubbert's original analysis did not apply to renewable resources, their overexploitation can result in a Hubbert-like peak. This has led to the concept of peak water.

Forest Resources

Beech forest – Grib Skov, Denmark

Forests are overexploited when they are logged at a rate faster than reforestation takes place. Reforestation competes with other land uses such as food production, livestock grazing, and living space for further economic growth. Historically utilization of forest products, including timber and fuel wood, have played a key role in human societies, comparable to the roles of water and cultivable land. Today, developed countries continue to utilize timber for building houses, and wood pulp for paper. In developing countries almost three billion people rely on wood for heating and cooking. Short-term economic gains made by conversion of forest to agriculture, or overexploitation of wood products, typically leads to loss of long-term income and long term biological productivity. West Africa, Madagascar, Southeast Asia and many other regions have experienced lower revenue because of overexploitation and the consequent declining timber harvests.

Biodiversity

Overexploitation is one of the main threats to global biodiversity. Other threats include pollution, introduced and invasive species, habitat fragmentation, habitat destruction, uncontrolled hybridization, global warming, ocean acidification and the driver behind many of these, human overpopulation.

The rich diversity of marine life inhabiting coral reefs attracts bioprospectors. Many coral reefs are overexploited; threats include coral mining, cyanide and blast fishing, and overfishing in general.

One of the key health issues associated with biodiversity is drug discovery and the availability of medicinal resources. A significant proportion of drugs are natural products derived, directly or indirectly, from biological sources. Marine ecosystems are of particular interest in this regard. However unregulated and inappropriate bioprospecting could potentially lead to overexploitation, ecosystem degradation and loss of biodiversity.

Endangered Species

Overexploitation threatens one-third of endangered vertebrates, as well as other groups. Excluding edible fish, the illegal trade in wildlife is valued at $10 billion per year. Industries responsible for this include the trade in bushmeat, the trade in Chinese medicine, and the fur trade. The Convention for International Trade in Endangered Species of Wild Fauna and Flora, or CITES was set up in order to control and regulate the trade in endangered animals. It currently protects, to a varying degree, some 33,000 species of animals and plants. It is estimated that a quarter of the endangered vertebrates in the United States of America and half of the endangered mammals is attributed to overexploitation.

It is not just humans that overexploit their resources. Overgrazing can occur naturally, caused by native fauna, as shown in the upper right.

All living organisms require resources to survive. Overexploitation of these resources for protracted periods can deplete natural stocks to the point where they are unable to recover within a short time frame. Humans have always harvested food and other resources they have needed to survive. Human populations, historically, were small, and methods of collection limited to small quantities. With an exponential increase in human population, expanding markets and increasing demand, combined with improved access and techniques for capture, are causing the exploitation of many species beyond sustainable levels. In practical terms, if continued, it reduces valuable resources to such low levels that their exploitation is no longer sustainable and can lead to the extinction of a species, in addition to having dramatic, unforeseen effects, on the ecosystem. Overexploitation often occurs rapidly as markets open, utilising previously untapped resources, or locally used species.

The Carolina parakeet was hunted to extinction.

Today, overexploitation and misuse of natural resources is an ever present threat for species richness. This is more prevalent when looking at island ecology and the species that inhabit them, as islands can be viewed as the world in miniature. Island endemic populations are more prone to extinction from overexploitation, as they often exist at low densities with reduced reproductive rates. A good example of this are island snails, such as the Hawaiian *Achatinella* and the French Polynesian *Partula*. Achatinelline snails have 15 species listed as extinct and 24 critically endangered while 60 species of partulidae are considered extinct with 14 listed as critically endangered. The WCMC have attributed over-collecting and very low lifetime fecundity for the extreme vulnerability exhibited among these species.

As another example, when the humble hedgehog was introduced to the Scottish island of Uist, the population greatly expanded and took to consuming and overexploiting shorebird eggs, with drastic consequences for their breeding success. Twelve species of avifauna are affected, with some species numbers being reduced by 39%.

Where there is substantial human migration, civil unrest, or war, controls may no longer exist. With civil unrest, for example in the Congo and Rwanda, firearms have become common and the breakdown of food distribution networks in such countries leaves the resources of the natural environment vulnerable. Animals are even killed as target practice, or simply to spite the government. Populations of large primates, such as gorillas and chimpanzees, ungulates and other mammals, may be reduced by 80% or more by hunting, and certain species may be eliminated altogether. This decline has been called the bushmeat crisis.

Overall, 50 bird species that have become extinct since 1500 (approximately 40% of the total) have been subject to overexploitation, including:

- Great Auk – The penguin-like bird of the north, hunted for its feathers, meat, fat and oil.

- Carolina parakeet – The only parrot species native to the eastern United States, was hunted for crop protection and its feathers.

Other species affected by overexploitation include:

- The international trade in fur: chinchilla, vicuña, giant otter and numerous cat species.

- Insect collectors: butterflies

- Horticulturists: New Zealand mistletoe (*Trilepidia adamsii*), orchids, cacti and many other plant species.

- Shell collectors: Marine molluscs

- Aquarium hobbyists: tropical fish

- Chinese medicine: bears, tigers, rhinos, seahorses, Asian black bear and saiga antelope

- Novelty pets: snakes, parrots, primates and big cats

Cascade Effects

Overexploitation of species can result in knock-on or cascade effects. This can particularly apply if, through overexploitation, a habitat loses its apex predator. Because of the loss of the top predator, a dramatic increase in their prey species can occur. In turn, the unchecked prey can then overexploit their own food resources until population numbers dwindle, possibly to the point of extinction.

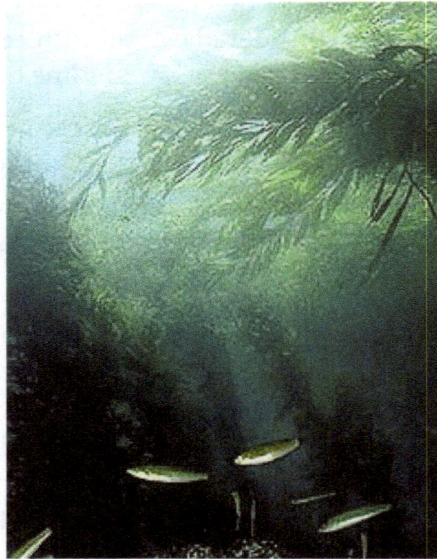

Overexploiting sea otters resulted in cascade effects which destroyed kelp forest ecosystems

A classic example of cascade effects occurred with sea otters. Starting before the 17th century and not phased out until 1911, sea otters were hunted aggressively for their exceptionally warm and valuable pelts, which could fetch up to $2500 US. This caused cascade effects through the kelp forest ecosystems along the Pacific Coast of North America.

One of the sea otters' primary food sources is the sea urchin. When hunters caused sea otter populations to decline, an ecological release of sea urchin populations occurred. The sea urchins then overexploited their main food source, kelp, creating urchin barrens, areas of seabed denuded of kelp, but carpeted with urchins. No longer having food to eat, the sea urchin became locally extinct as well. Also, since kelp forest ecosystems are homes to many other species, the loss of the kelp caused other cascade effects of secondary extinctions.

In 1911, when only one small group of 32 sea otters survived in a remote cove, an international treaty was signed to prevent further exploitation of the sea otters. Under heavy protection, the otters multiplied and repopulated the depleted areas, which slowly recovered. More recently, with declining numbers of fish stocks, again due to overexploitation, killer whales have experienced a food shortage and have been observed feeding on sea otters, again reducing their numbers.

Human Overpopulation

Human overpopulation occurs if the number of people in a group exceeds the carrying capacity of the region occupied by that group. Overpopulation can further be viewed, in a long term perspective, as existing when a population cannot be maintained given the rapid depletion of non-renewable resources or given the degradation of the capacity of the environment to give support to the population.

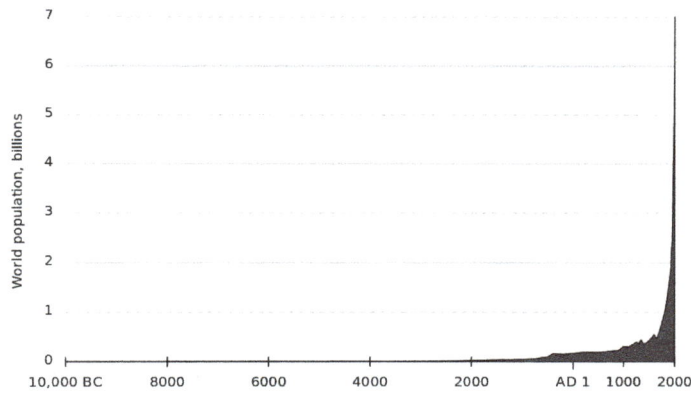

Graph of human population from 10,000 BCE to 2000 CE. It shows the extremely rapid growth in the world population that has taken place since the eighteenth century.

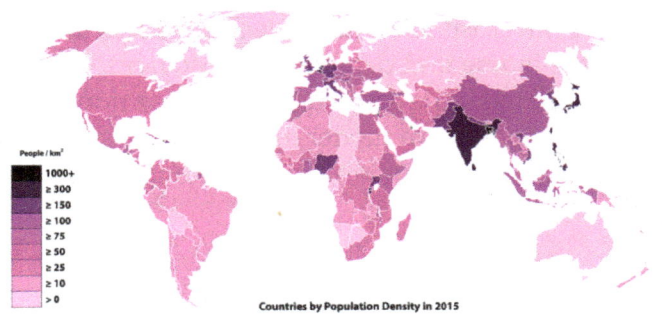

Map of population density by country, per square kilometer.

Areas of high population densities, calculated in 1994

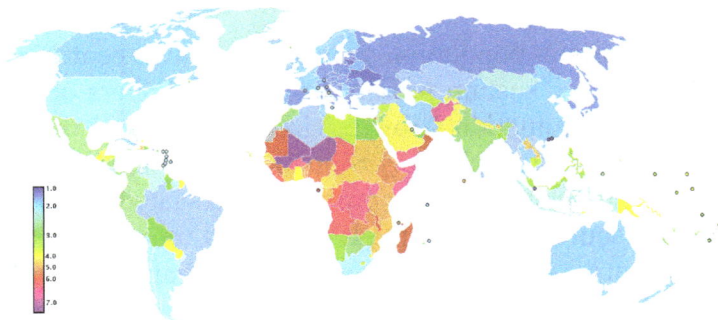

Map of countries and territories by fertility rate

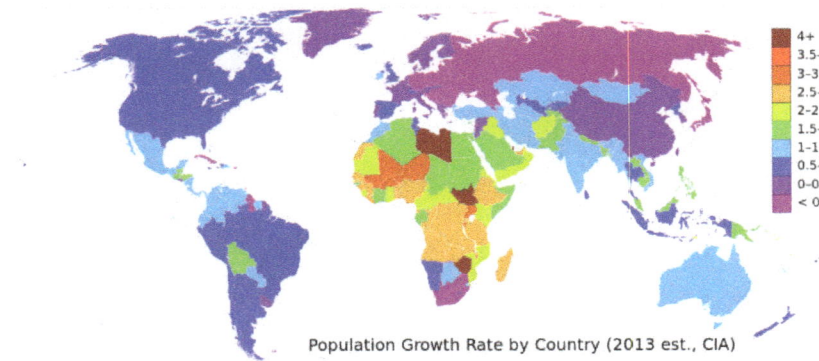

Population Growth Rate by Country (2013 est., CIA)

Human population growth rate in percent, with the variables of births, deaths, immigration, and emigration – 2013

The term *human overpopulation* often refers to the relationship between the entire human population and its environment: the Earth, or to smaller geographical areas such as countries. Overpopulation can result from an increase in births, a decline in mortality rates, an increase in immigration, or an unsustainable biome and depletion of resources. It is possible for very sparsely populated areas to be overpopulated if the area has a meagre or non-existent capability to sustain life (e.g. a desert). Advocates of population moderation cite issues like quality of life, carrying capacity and risk of starvation as a basis to argue against continuing high human population growth and for population decline.

Overview

Human population has been growing continuously since the end of the Black Death, around the year 1350, although the most significant increase has been in the last 50 years, mainly due to medical advancements and increase in agricultural productivity. The rate of population growth has been declining since the 1980s. The United Nations has expressed concern on continued excessive population growth in sub-Saharan Africa. Recent research has demonstrated that those concerns are well grounded. As of September 27, 2016 the world's human population is estimated to be 7.354 billion by the United States Census Bureau, and over 7 billion by the United Nations. Most contemporary estimates for the carrying capacity of the Earth under existing conditions are between 4 billion and 16 billion. Depending on which estimate is used, human overpopulation may or may not have already occurred. Nevertheless, the rapid recent increase in human population is causing some concern. The population is expected to reach between 8 and 10.5 billion between the years 2040 and 2050. In May 2011, the United Nations increased the medium variant projections to 9.3 billion for 2050 and 10.1 billion for 2100.

The recent rapid increase in human population over the past three centuries has raised concerns that the planet may not be able to sustain present or future numbers of inhabitants. The Inter-Academy Panel Statement on Population Growth, circa 1994, stated that many environmental problems, such as rising levels of atmospheric carbon dioxide, global warming, and pollution, are aggravated by the population expansion. Other problems associated with overpopulation include the increased demand for resources such as fresh water and food, starvation and malnutrition, consumption of natural resources (such as fossil fuels) faster than the rate of regeneration, and a deterioration in living conditions. Wealthy but highly populated territories like Britain rely on food imports from overseas. This was severely felt during the World Wars when, despite food efficiency initiatives like "dig for victory" and food rationing, Britain needed to fight to secure import routes. However, many believe that waste and over-consumption, especially by wealthy nations, is putting more strain on the environment than overpopulation.

Most countries have no direct policy of limiting their birth rates, but the rates have still fallen due to education about family planning and increasing access to birth control and contraception. Only China has imposed legal restrictions on having more than one child. Extraterrestrial settlement and other technical solutions have been proposed as ways to mitigate overpopulation in the future.

History of Concern

Concern about overpopulation is an ancient topic. Tertullian was a resident of the city of Carthage in the second century CE, when the population of the world was about 190 million (only 3-4% of what it is today). He notably said: "What most frequently meets our view (and occasions complaint) is our teeming population. Our numbers are burdensome to the world, which can hardly support us…. In very deed, pestilence, and famine, and wars, and earthquakes have to be regarded as a remedy for nations, as the means of pruning the luxuriance of the human race." Before that, Plato, Aristotle and others broached the topic as well.

Thousands of scooters make their way through Ho Chi Minh City, Vietnam.

Throughout history, population growth has usually been slow despite high birth rates, due to war, plagues and other diseases, and high infant mortality. During the 750 years before the Industrial Revolution, the world's population increased very slowly, remaining under 250 million.

By the beginning of the 19th century, the world population had grown to a billion individuals, and intellectuals such as Thomas Malthus predicted that mankind would outgrow its available resources, because a finite amount of land would be incapable of supporting a population with a limitless potential for increase. Mercantillists argued that a large population was a form of wealth, which made it possible to create bigger markets and armies.

During the 19th century, Malthus's work was often interpreted in a way that blamed the poor alone for their condition and helping them was said to worsen conditions in the long run. This resulted, for example, in the English poor laws of 1834 and in a hesitating response to the Irish Great Famine of 1845–52.

The UN Population Assessment Report of 2003 projects that the world population will plateau by 2050 and will remain stable until 2300. Alex Berezow, editor of *RealClearScience*, states that overpopulation is not a Western world problem, and people often cite China and India as major population contributors; however, he notes that with rising wealth in those countries, population growth will begin to slow, and this is a proven factor in the economic stability of nations.

Human Population

History of Population Growth

The human population has gone through a number of periods of growth since the dawn of civilization in the Holocene period, around 10,000 BCE. The beginning of civilization roughly coincides with the receding of glacial ice following the end of the last glacial period. It is estimated that between 1-5 million people, subsisting on hunting and foraging, inhabited the Earth in the period before the Neolithic Revolution, when human activity shifted away from hunter-gathering and towards very primitive farming.

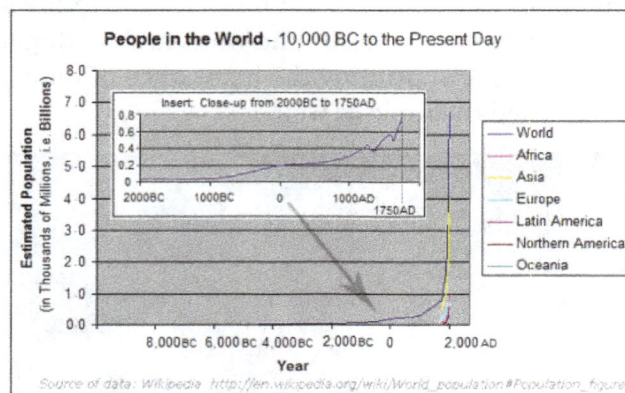

Data from World Population.

Population	
Year	**Billion**
1804	1
1927	2
1959	3
1974	4
1987	5
1999	6
2011	7
2020	7.7 (estimate)

Around 8000 BCE, at the dawn of agriculture, the population of the world was approximately 5 million. The next several millennia saw a steady increase in the population, with very rapid growth beginning in 1000 BCE, and a peak of between 200 and 300 million people in 1 BCE.

The Plague of Justinian caused Europe's population to drop by around 50% between 541 and the 8th century. Steady growth resumed in 800 CE. However, growth was again disrupted by frequent plagues; most notably, the Black Death during the 14th century. The effects of the Black Death are thought to have reduced the world's population, then at an estimated 450 million, to between 350 and 375 million by 1400. The population of Europe stood at over 70 million in 1340; these levels did not return until 200 years later. England's population reached an estimated 5.6 million in 1650, up from an estimated 2.6 million in 1500. New crops from the Americas via the Spanish colonizers in the 16th century contributed to the population growth.

In other parts of the globe, China's population at the founding of the Ming dynasty in 1368 stood close to 60 million, approaching 150 million by the end of the dynasty in 1644. The population of the Americas in 1500 may have been between 50 and 100 million.

Encounters between European explorers and populations in the rest of the world often introduced local epidemics of extraordinary virulence. Archaeological evidence indicates that the death of around 90% of the Native American population of the New World was caused by Old World diseases such as smallpox, measles, and influenza. Europeans introduced diseases alien to the indigenous people, therefore they did not have immunity to these foreign diseases.

After the start of the Industrial Revolution, during the 18th century, the rate of population growth began to increase. By the end of the century, the world's population was estimated at just under 1 billion. At the turn of the 20th century, the world's population was roughly 1.6 billion. By 1940, this figure had increased to 2.3 billion.

Population growth 1990–2009 (%)	
World	28.4%
Africa	58.4%
Middle East	53.4%
Asia (except China)	36.9%
Latin America	32.0%
OECD North America	25.1%
China	17.3%
OECD Europe	9.9%
OECD Pacific	9.5%
Non-OECD Europe and Eurasia	-2.7%

Dramatic growth beginning in 1950 (above 1.8% per year) coincided with greatly increased food production as a result of the industrialization of agriculture brought about by the Green Revolution. The rate of human population growth peaked in 1964, at about 2.1% per year. For example, Indonesia's population grew from 97 million in 1961 to 237.6 million in 2010, a 145% increase in 49 years. In India, the population grew from 361.1 million people in 1951 to just over 1.2 billion by 2011, a 235% increase in 60 years.

Continent	1900 population
Africa	133 million
Asia	904 million
Europe	408 million
Latin America and Caribbean	74 million
North America	82 million

There is concern over the sharp population increase in many countries, especially in Sub-Saharan Africa, that has occurred over the last several decades, and that it is creating problems with land management, natural resources and access to water supplies.

The population of Chad has, for example, grown from 6,279,921 in 1993 to 10,329,208 in 2009. Niger, Uganda, Nigeria, Tanzania, Ethiopia and the DRC are witnessing a similar growth in population. The situation is most acute in western, central and eastern Africa. Refugees from places like Sudan have further strained the resources of neighboring states like Chad and Egypt. Chad is also host to roughly 255,000 refugees from Sudan's Darfur region, and about 77,000 refugees from the Central African Republic, while approximately 188,000 Chadians have been displaced by their own civil war and famines, have either fled to either the Sudan, the Niger or, more recently, Libya.

Projections of Population Growth

Continent	Projected 2050 population
Africa	1.8 billion
Asia	5.3 billion
Europe	628 million
Latin America and Caribbean	809 million
North America	392 million

According to projections, the world population will continue to grow until at least 2050, with the population reaching 9 billion in 2040, and some predictions putting the population as high as 11 billion in 2050. By 2100, the population could reach 15 billion. Walter Greiling projected in the 1950s that world population would reach a peak of about nine billion, in the 21st century, and then stop growing, after a readjustment of the Third World and a sanitation of the tropics.

According to the United Nations' World Population Prospects report:

- The world population is currently growing by approximately 74 million people per year. Current United Nations predictions estimate that the world population will reach 9.0 billion around 2050, assuming a decrease in average fertility rate from 2.5 down to 2.0.

- Almost all growth will take place in the less developed regions, where today's 5.3 billion population of underdeveloped countries is expected to increase to 7.8 billion in 2050. By contrast, the population of the more developed regions will remain mostly unchanged, at 1.2 billion. An exception is the United States population, which is expected to increase by 44% from 2008 to 2050.

- In 2000–2005, the average world fertility was 2.65 children per woman, about half the level in 1950–1955 (5 children per woman). In the medium variant, global fertility is projected to decline further to 2.05 children per woman.

- During 2005–2050, nine countries are expected to account for half of the world's projected population increase: India, Pakistan, Nigeria, Democratic Republic of the Congo, Bangladesh, Uganda, United States, Ethiopia, and China, listed according to the size of their contribution to population growth. China would be higher still in this list were it not for its one-child policy.

- Global life expectancy at birth is expected to continue rising from 65 years in 2000–2005 to 75 years in 2045–2050. In the more developed regions, the projection is to 82 years by 2050. Among the least developed countries, where life expectancy today is just under 50 years, it is expected to increase to 66 years by 2045–2050.

- The population of 51 countries or areas is expected to be lower in 2050 than in 2005.

- During 2005–2050, the net number of international migrants to more developed regions is projected to be 98 million. Because deaths are projected to exceed births in the more developed regions by 73 million during 2005–2050, population growth in those regions will largely be due to international migration.

- In 2000–2005, net migration in 28 countries either prevented population decline or doubled at least the contribution of natural increase (births minus deaths) to population growth.

- Birth rates are now falling in a small percentage of developing countries, while the actual populations in many developed countries would fall without immigration.

Urban Growth

In 1800 only 3% of the world's population lived in cities. By the 20th century's close, 47% did so. In 1950 there were 83 cities with populations exceeding one million; but by 2007 this had risen to 468 agglomerations of more than one million. If the trend continues, the world's urban population will double every 38 years, according to researchers. The UN forecasts that today's urban population of 3.2 billion will rise to nearly 5 billion by 2030, when three out of five people will live in cities.

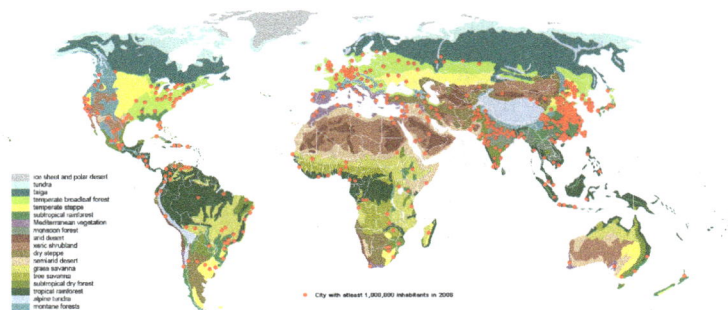

Urban areas with at least one million inhabitants in 2006. 3% of the world's population lived in cities in 1800, rising to 47% at the end of the twentieth century.

The increase will be most dramatic in the poorest and least-urbanised continents, Asia and Africa. Projections indicate that most urban growth over the next 25 years will be in developing countries. One billion people, one-seventh of the world's population, or one-third of urban population, now live in shanty towns, which are seen as "breeding grounds" for social problems such as crime, drug addiction, alcoholism, poverty and unemployment. In many poor countries, slums exhibit high rates of disease due to unsanitary conditions, malnutrition, and lack of basic health care.

In 2000, there were 18 megacities – conurbations such as Tokyo, Beijing, Guangzhou, Seoul, Karachi, Mexico City, Mumbai, São Paulo and New York City – that have populations in excess of 10 million inhabitants. Greater Tokyo already has 35 million, more than the entire population of Canada (at 34.1 million).

According to the *Far Eastern Economic Review*, Asia alone will have at least 10 'hypercities' by 2025, that is, cities inhabited by more than 19 million people, including Jakarta (24.9 million people), Dhaka (25 million), Karachi (26.5 million), Shanghai (27 million) and Mumbai (33 million). Lagos has grown from 300,000 in 1950 to an estimated 15 million today, and the Nigerian government estimates that city will have expanded to 25 million residents by 2015. Chinese experts forecast that Chinese cities will contain 800 million people by 2020.

Causes

From a historical perspective, technological revolutions have coincided with population explosions. There have been three major technological revolutions – the tool-making revolution, the agricultural revolution, and the industrial revolution – all of which allowed humans more access to food, resulting in subsequent population explosions. For example, the use of tools, such as bow and arrow, allowed primitive hunters greater access to high energy foods (e.g. animal meat). Similarly, the transition to farming about 10,000 years ago greatly increased the overall food supply, which was used to support more people. Food production further increased with the industrial revolution as machinery, fertilizers, herbicides, and pesticides were used to increase land under cultivation as well as crop yields. In short, similar to bacteria that multiply in response to increased food supply, humans have increased their population as soon as food became more abundant as a result of technological innovations.

Street in Kathmandu

Significant increases in human population occur whenever the birth rate exceeds the death rate for extended periods of time. Traditionally, the fertility rate is strongly influenced by cultural and social norms that are rather stable and therefore slow to adapt to changes in the social, technological, or environmental conditions. For example, when death rates fell during the 19th and 20th century – as a result of improved sanitation, child immunizations, and other advances in medicine – allowing more newborns to survive, the fertility rate did not adjust downward fast enough, resulting in significant population growth. Prior to these changes, seven out of ten children died before reaching reproductive age, while today about 95% of newborns in industrialized nations reach adulthood.

Agriculture

Agriculture has been the main factor behind human population growth. This dates back to prehistoric times, when agricultural methods were first developed, and continues to the present day, with fertilizers, agrochemicals, large-scale mechanization, genetic manipulation, and other technologies.

Psychological Factors

Human psychology and the cycle of entrenched poverty, as well as the rest of the world's reaction to it, are also causative factors. Areas with greater burden of disease and warfare, contrary to popular belief, do not experience less population growth over the long term, but far more over a sustained period as poverty becomes further entrenched

This is because parents and siblings who have experienced calamitous conditions suffer from a kind of post traumatic stress syndrome about losing their family members and overcompensate by having "extra" babies. These extra babies and calamities fuel a vicious cycle, and only in the small minority of cases does it cease

As this cycle is compounded over generations, calamities such as disaster or war take on a multiplier effect. For example, the AIDS crisis in Africa is said to have killed 30 million to date, yet during the last two decades money and initiatives to lower population growth by contraception have been sidelined in favor of combating HIV, feeding the population explosion that we see in Africa today. In 1990, this continent's population was roughly 600 million; today it is over 1,050 million, 150 million more than if the HIV/AIDS crisis had never occurred.

Extremes

Population growth rates between 1950 and 2012 range from a 0.5% increase in the case of Bulgaria to a more than 100 fold increase for the United Arab Emirates (from 79,050 to 8.5 million). Roughly half of all nations have quadrupled their populations since 1950.

Demographic Transition

The theory of demographic transition held that, after the standard of living and life expectancy increase, family sizes and birth rates decline. However, as new data has become available, it has been observed that after a certain level of development (HDI equal to 0.86 or higher) the fertility increases again. This means that both the worry that the theory generated about aging populations and the complacency it bred regarding the future environmental impact of population growth are misguided.

Factors cited in the old theory included such social factors as later ages of marriage, the growing desire of many women in such settings to seek careers outside child rearing and domestic work, and the decreased need of children in industrialized settings. The latter factor stems from the fact that children perform a great deal of work in small-scale agricultural societies, and work less in industrial ones; it has been cited to explain the decline in birth rates in industrializing regions.

Many countries have high population growth rates but lower total fertility rates because high population growth in the past skewed the age demographic toward a young age, so the population still rises as the more numerous younger generation approaches maturity.

"Demographic entrapment" is a concept developed by Maurice King, Honorary Research Fellow at the University of Leeds, who posits that this phenomenon occurs when a country has a population larger than its carrying capacity, no possibility of migration, and exports too little to be able to import food. This will cause starvation. He claims that for example many sub-Saharan nations are or will become stuck in demographic entrapment, instead of having a demographic transition.

For the world as a whole, the number of children born per woman decreased from 5.02 to 2.65 between 1950 and 2005. A breakdown by region is as follows:

- Europe – 2.66 to 1.41

- North America – 3.47 to 1.99

- Oceania – 3.87 to 2.30

- Central America – 6.38 to 2.66

- South America – 5.75 to 2.49

- Asia (excluding Middle East) – 5.85 to 2.43

- Middle East & North Africa – 6.99 to 3.37

- Sub-Saharan Africa – 6.7 to 5.53

Excluding the observed reversal in fertility decrease for high development, the projected world number of children born per woman for 2050 would be around 2.05. Only the Middle East & North Africa (2.09) and Sub-Saharan Africa (2.61) would then have numbers greater than 2.05.

Carrying Capacity

Some groups (for example, the World Wide Fund for Nature and Global Footprint Network) have stated that the carrying capacity for the human population has been exceeded as measured using the Ecological Footprint. In 2006, WWF's "Living Planet Report" stated that in order for all humans to live with the current consumption patterns of Europeans, we would be spending three times more than what the planet can renew. Humanity as a whole was using, by 2006, 40 percent more than what Earth can regenerate. However, Roger Martin of Population Matters states the view: "the poor want to get rich, and I want them to get rich," with a later addition, "of course we have to change consumption habits,... but we've also got to stabilise our numbers".

A family planning placard in Ethiopia. It shows some negative effects of having too many children.

But critics question the simplifications and statistical methods used in calculating Ecological Footprints. Therefore, Global Footprint Network and its partner organizations have engaged with national governments and international agencies to test the results – reviews have been produced by France, Germany, the European Commission, Switzerland, Luxembourg, Japan and the United Arab Emirates. Some point out that a more refined method of assessing Ecological Footprint is to designate sustainable versus non-sustainable categories of consumption. However, if yield estimates were adjusted fTor sustainable levels of production, the yield figures would be lower, and hence the overshoot estimated by the Ecological Footprint method even higher.

Other studies give particular attention to resource depletion and increased world affluence.

In a 1994 study titled *Food, Land, Population and the U.S. Economy*, David Pimentel and Mario Giampietro estimated the maximum U.S. population for a sustainable economy at 200 million. And in order to achieve a sustainable economy and avert disaster, the United States would have to reduce its population by at least one-third, and world population would have to be reduced by two-thirds.

Many quantitative studies have estimated the world's carrying capacity for humans, that is, a limit to the world population. A meta-analysis of 69 such studies suggests a point estimate of the limit to be 7.7 billion people, while lower and upper meta-bounds for current technology are estimated as 0.65 and 98 billion people, respectively. They conclude: "recent predictions of stabilized world population levels for 2050 exceed several of our meta-estimates of a world population limit".

Effects of Human Overpopulation

Raw numbers of people are only one factor in the effects of people. The lifestyle (including overall affluence and resource utilization) and the pollution (including carbon footprint) are equally important. In 2008, the New York Times stated that the inhabitants of the developed nations of the world consume resources like oil and metals at a rate almost 32 times greater than those of the developing world, who make up the majority of the human population.

Some problems associated with or exacerbated by human overpopulation and over-consumption are:

- Inadequate fresh water for drinking as well as sewage treatment and effluent discharge. Some countries, like Saudi Arabia, use energy-expensive desalination to solve the problem of water shortages.

- Depletion of natural resources, especially fossil fuels.

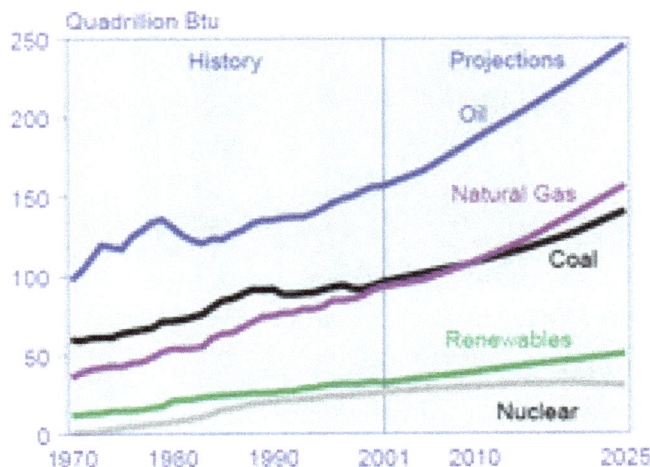

World energy consumption & predictions, 1970–2025.

- Increased levels of air pollution, water pollution, soil contamination and noise pollution. Once a country has industrialized and become wealthy, a combination of government regulation and technological innovation causes pollution to decline substantially, even as the population continues to grow.

- Deforestation and loss of ecosystems that valuably contribute to the global atmospheric oxygen and carbon dioxide balance; about eight million hectares of forest are lost each year.

- Changes in atmospheric composition and consequent global warming.

- Loss of arable land and increase in desertification. Deforestation and desertification can be reversed by adopting property rights, and this policy is successful even while the human population continues to grow.

- Mass species extinctions from reduced habitat in tropical forests due to slash-and-burn techniques that sometimes are practiced by shifting cultivators, especially in countries with rapidly expanding rural populations; present extinction rates may be as high as 140,000 species lost per year. As of February 2011, the IUCN Red List lists a total of 801 animal species having gone extinct during recorded human history.

- High infant and child mortality. High rates of infant mortality are associated with poverty. Rich countries with high population densities have low rates of infant mortality.

- Intensive factory farming to support large populations. It results in human threats including the evolution and spread of antibiotic resistant bacteria diseases, excessive air and water pollution, and new viruses that infect humans.

A child suffering extreme malnutrition in India, 1972

- Increased chance of the emergence of new epidemics and pandemics. For many environmental and social reasons, including overcrowded living conditions, malnutrition and inadequate, inaccessible, or non-existent health care, the poor are more likely to be exposed to infectious diseases.

- Starvation, malnutrition or poor diet with ill health and diet-deficiency diseases (e.g. rickets). However, rich countries with high population densities do not have famine.

- Poverty coupled with inflation in some regions and a resulting low level of capital formation. Poverty and inflation are aggravated by bad government and bad economic policies. Many countries with high population densities have eliminated absolute poverty and keep their inflation rates very low.

- Low life expectancy in countries with fastest growing populations.

- Unhygienic living conditions for many based upon water resource depletion, discharge of raw sewage and solid waste disposal. However, this problem can be reduced with the adoption of sewers. For example, after Karachi, Pakistan installed sewers, its infant mortality rate fell substantially.

- Elevated crime rate due to drug cartels and increased theft by people stealing resources to survive.

- Conflict over scarce resources and crowding, leading to increased levels of warfare.

- Less personal freedom and more restrictive laws. Laws regulate interactions between humans. Law "serves as a primary social mediator of relations between people". The higher the population density, the more frequent such interactions become, and thus there develops a need for more laws and/or more restrictive laws to regulate these interactions. It was even speculated by Aldous Huxley in 1958 that democracy is threatened due to overpopulation, and could give rise to totalitarian style governments.

- David Attenborough described the level of human population on the planet as a multiplier of all other environmental problems.

Many of these problems are explored in the dystopic science fiction film *Soylent Green*, where an overpopulated Earth suffers from food shortages, depleted resources and poverty and in the documentary "Aftermath: Population Overload".

Some economists, such as Thomas Sowell and Walter E. Williams argue that third world poverty and famine are caused in part by bad government and bad economic policies. Most biologists and sociologists see overpopulation as a serious threat to the quality of human life.

Resources

Overpopulation does not depend only on the size or density of the population, but on the ratio of population to available sustainable resources. It also depends on how resources are managed and distributed throughout the population.

The resources to be considered when evaluating whether an ecological niche is overpopulated include clean water, clean air, food, shelter, warmth, and other resources necessary to sustain life. If the quality of human life is addressed, there may be additional resources considered, such as medical care, education, proper sewage treatment, waste disposal and energy supplies. Overpopulation places competitive stress on the basic life sustaining resources, leading to a diminished quality of life.

Directly related to maintaining the health of the human population is water supply, and it is one of the resources that experience the biggest strain. With the global population at about 7.125 billion, and each human theoretically needing 2 liters of drinking water, there is a demand for 14.25 billion liters of water each day to meet the minimum requirement for healthy living (United). Weather patterns, elevation, and climate all contribute to uneven distribution of fresh drinking water. Without clean water, good health is not a viable option. Besides drinking, water is used to create sanitary living conditions and is the basis of creating a healthy environment fit to hold human life. In addition to drinking water, water is also used for bathing, washing clothes and dishes, flushing toilets, a variety of cleaning methods, recreation, watering lawns, and farm irrigation. Irrigation poses one of the largest problems, because without sufficient water to irrigate crops, the crops die and then there is the problem of food rations and starvation. In addition to water needed for crops and food, there is limited land area dedicated to food production, and not much more that is suitable to be added. Arable land, needed to sustain the growing population, is also a factor because land being under or over cultivated easily upsets the delicate balance of nutrition supply. There are also problems with location of arable land with regard to proximity to countries and relative population (Bashford 240). Access to nutrition is an important limiting factor in population sustainability and growth. No increase in arable land added to the still increasing human population will eventually pose a serious conflict. Only 38% of the land area of the globe is dedicated to agriculture, and there is not room for much more. Although plants produce 54 billion metric tons of carbohydrates per year, when the population is expected to grow to 9 billion by 2050, the plants may not be able to keep up (Biello). Food supply is a primary example of how a resource reacts when its carrying capacity is exceeded. By trying to grow more and more crops off of the same amount of land, the soil becomes exhausted. Because the soil is exhausted, it is then unable to produce the same amount of food as before, and is overall less productive. Therefore, by using resources beyond a sustainable level, the resource become nullified and ineffective, which further increases the disparity between the demand for a resource and the availability of a resource. There must be a shift to provide adequate recovery time to each one of the supplies in demand to support contemporary human lifestyles.

An industrial area, with a power plant, south of Yangzhou's downtown, China

David Pimentel has stated that "With the imbalance growing between population numbers and vital life sustaining resources, humans must actively conserve cropland, freshwater, energy, and biological resources. There is a need to develop renewable energy resources. Humans everywhere must understand that rapid population growth damages the Earth's resources and diminishes human well-being."

These reflect the comments also of the United States Geological Survey in their paper The Future of Planet Earth: Scientific Challenges in the Coming Century. "As the global population continues to grow...people will place greater and greater demands on the resources of our planet, including mineral and energy resources, open space, water, and plant and animal resources." "Earth's natural wealth: an audit" by New Scientist magazine states that many of the minerals that we use for a variety of products are in danger of running out in the near future. A handful of geologists around the world have calculated the costs of new technologies in terms of the materials they use and the implications of their spreading to the developing world. All agree that the planet's booming population and rising standards of living are set to put unprecedented demands on the materials that only Earth itself can provide. Limitations on how much of these materials is available could even mean that some technologies are not worth pursuing long term.... "Virgin stocks of several metals appear inadequate to sustain the modern 'developed world' quality of life for all of Earth's people under contemporary technology".

On the other hand, some cornucopian researchers, such as Julian L. Simon and Bjørn Lomborg believe that resources exist for further population growth. In a 2010 study, they concluded that "there are not (and will never be) too many people for the planet to feed" according to The Independent. Some critics warn, this will be at a high cost to the Earth: "the technological optimists are probably correct in claiming that overall world food production can be increased substantially over the next few decades...[however] the environmental cost of what Paul R. and Anne H. Ehrlich describe as 'turning the Earth into a giant human feedlot' could be severe. A large expansion of agriculture to provide growing populations with improved diets is likely to lead to further deforestation, loss of species, soil erosion, and pollution from pesticides and fertilizer runoff as farming intensifies and new land is brought into production." Since we are intimately dependent upon the living systems of the Earth, some scientists have questioned the wisdom of further expansion.

According to the Millennium Ecosystem Assessment, a four-year research effort by 1,360 of the world's prominent scientists commissioned to measure the actual value of natural resources to humans and the world, "The structure of the world's ecosystems changed more rapidly in the second half of the twentieth century than at any time in recorded human history, and virtually all of Earth's ecosystems have now been significantly transformed through human actions." "Ecosystem services, particularly food production, timber and fisheries, are important for employment and economic activity. Intensive use of ecosystems often produces the greatest short-term advantage, but excessive and unsustainable use can lead to losses in the long term. A country could cut its forests and deplete its fisheries, and this would show only as a positive gain to GDP, despite the loss of capital assets. If the full economic value of ecosystems were taken into account in decision-making, their degradation could be significantly slowed down or even reversed."

Another study by the United Nations Environment Programme (UNEP) called the Global Environment Outlook which involved 1,400 scientists and took five years to prepare comes to similar conclusions. It "found that human consumption had far outstripped available resources. Each person on Earth now requires a third more land to supply his or her needs than the planet can supply." It faults a failure to "respond to or recognize the magnitude of the challenges facing the people and the environment of the planet... 'The systematic destruction of the Earth's natural and nature-based resources has reached a point where the economic viability of economies is being challenged – and where the bill we hand to our children may prove impossible to pay'... The report's authors say its objective is 'not to present a dark and gloomy scenario, but an urgent call to action'. It warns that tackling the problems may affect the vested interests of powerful groups, and that the environment must be moved to the core of decision-making... '

Although all resources, whether mineral or other, are limited on the planet, there is a degree of self-correction whenever a scarcity or high-demand for a particular kind is experienced. For example, in 1990 known reserves of many natural resources were higher, and their prices lower, than in 1970, despite higher demand and higher consumption. Whenever a price spike would occur, the market tended to correct itself whether by substituting an equivalent resource or switching to a new technology.

Fresh Water

Fresh water supplies, on which agriculture depends, are running low worldwide. This water crisis is only expected to worsen as the population increases.

Potential problems with dependence on desalination are reviewed below, however, the majority of the world's freshwater supply is contained in the polar icecaps, and underground river systems accessible through springs and wells.

Fresh water can be obtained from salt water by desalination. For example, Malta derives two thirds of its freshwater by desalination. A number of nuclear powered desalination plants exist; However, the high costs of desalination, especially for poor countries, make impractical the transport of large amounts of desalinated seawater to interiors of large countries. The cost of desalination varies; Israel is now desalinating water for a cost of 53 cents per cubic meter, Singapore at 49 cents per cubic meter. In the United States, the cost is 81 cents per cubic meter ($3.06 for 1,000 gallons).

According to a 2004 study by Zhou and Tol, "one needs to lift the water by 2000 m, or transport it over more than 1600 km to get transport costs equal to the desalination costs. Desalinated water is expensive in places that are both somewhat far from the sea and somewhat high, such as Riyadh and Harare. In other places, the dominant cost is desalination, not transport. This leads to somewhat lower costs in places like Beijing, Bangkok, Zaragoza, Phoenix, and, of course, coastal cities like Tripoli." Thus while the study is generally positive about the technology for affluent areas that are proximate to oceans, it concludes that "Desalinated water may be a solution for some water-stress regions, but not for places that are poor, deep in the interior of a continent, or at high elevation. Unfortunately, that includes some of the places with biggest water problems." Another potential problem with desalination is the byproduction of saline brine, which can be a major cause of marine pollution when dumped back into the oceans at high temperatures."

The world's largest desalination plant is the Jebel Ali Desalination Plant (Phase 2) in the United Arab Emirates, which can produce 300 million cubic metres of water per year, or about 2500 gallons per second. The largest desalination plant in the US is the one at Tampa Bay, Florida, which began desalinating 25 million gallons (95000 m³) of water per day in December 2007. A 17 January 2008, article in the *Wall Street Journal* states, "Worldwide, 13,080 desalination plants produce more than 12 billion gallons of water a day, according to the International Desalination

Association." After being desalinated at Jubail, Saudi Arabia, water is pumped 200 miles (320 km) inland though a pipeline to the capital city of Riyadh.

However, new data originating from the GRACE experiments and isotopic testing done by the IAEA show that the Nubian aquifer—which is under the largest, driest part of the earth's surface, has enough water in it to provide for "at least several centuries". In addition to this, new and highly detailed maps of the earth's underground reservoirs will be soon created from these technologies that will further allow proper budgeting of cheap water.

Food

Some scientists argue that there is enough food to support the world population, and some dispute this, particularly if sustainability is taken into account.

Many countries rely heavily on imports. Egypt and Iran rely on imports for 40% of their grain supply. Yemen and Israel import more than 90%. And just 6 countries – Argentina, Australia, Canada, France, Thailand and the USA – supply 90% of grain exports. In recent decades the US alone supplied almost half of world grain exports.

A 2001 United Nations report says population growth is "the main force driving increases in agricultural demand" but "most recent expert assessments are cautiously optimistic about the ability of global food production to keep up with demand for the foreseeable future (that is to say, until approximately 2030 or 2050)", assuming declining population growth rates.

However, the observed figures for 2007 show an actual increase in absolute numbers of undernourished people in the world, 923 million in 2007 versus 832 million in 1995.; the more recent FAO estimates point to an even more dramatic increase, to 1.02 billion in 2009.

Global Perspective

The amounts of natural resources in this context are not necessarily fixed, and their distribution is not necessarily a zero-sum game. For example, due to the Green Revolution and the fact that more and more land is appropriated each year from wild lands for agricultural purposes, the worldwide production of food had steadily increased up until 1995. World food production per person was considerably higher in 2005 than 1961.

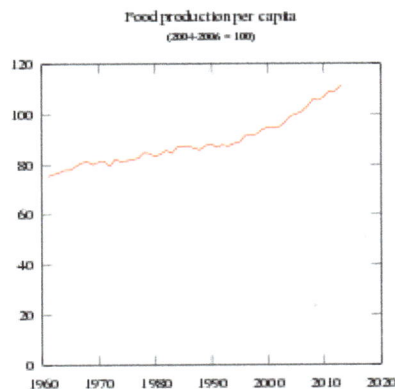

Growth in food production has been greater than population growth.

As world population doubled from 3 billion to 6 billion, daily calorie consumption in poor countries increased from 1,932 to 2,650, and the percentage of people in those countries who were malnourished fell from 45% to 18%. This suggests that Third World poverty and famine are caused by underdevelopment, not overpopulation. However, others question these statistics. From 1950 to 1984, as the Green Revolution transformed agriculture around the world, grain production increased by over 250%. The world population has grown by about four billion since the beginning of the Green Revolution and most believe that, without the Revolution, there would be greater famine and malnutrition than the UN presently documents.

The number of people who are overweight has surpassed the number who are undernourished. In a 2006 news story, MSNBC reported, "There are an estimated 800 million undernourished people and more than a billion considered overweight worldwide." The U.S. has one of the highest rates of obesity in the world. However, studies show that wealthy and educated people are far likelier to eat healthy food, indicating obesity is a disease related to poverty and lack of education and excessive advertising of unhealthy eatables at cheaper cost, high in calories, with little nutritive value are consumed.

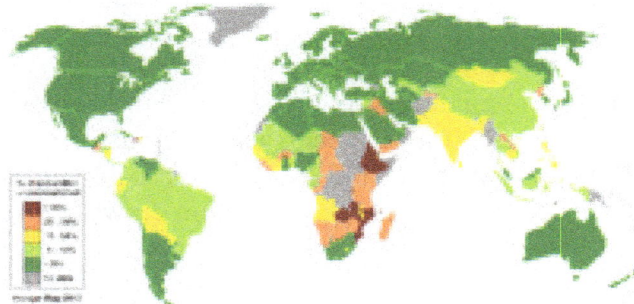

Percentage of population suffering from malnutrition by country, according to United Nations statistics.

The Food and Agriculture Organization of the United Nations states in its report *The State of Food Insecurity in the World 2006*, that while the number of undernourished people in the developing countries has declined by about three million, a smaller proportion of the populations of developing countries is undernourished today than in 1990–92: 17% against 20%. Furthermore, FAO's projections suggest that the proportion of hungry people in developing countries could be halved from 1990–92 levels to 10% by 2015. The FAO also states "We have emphasized first and foremost that reducing hunger is no longer a question of means in the hands of the global community. The world is richer today than it was ten years ago. There is more food available and still more could be produced without excessive upward pressure on prices. The knowledge and resources to reduce hunger are there. What is lacking is sufficient political will to mobilize those resources to the benefit of the hungry."

As of 2008, the price of grain has increased due to more farming used in biofuels, world oil prices at over $100 a barrel, global population growth, climate change, loss of agricultural land to residential and industrial development, and growing consumer demand in China and India Food riots have recently taken place in many countries across the world. An epidemic of stem rust on wheat caused by race Ug99 is currently spreading across Africa and into Asia and is causing major concern. A virulent wheat disease could destroy most of the world's main wheat crops, leaving millions to starve. The fungus has spread from Africa to Iran, and may already be in Afghanistan and Pakistan.

Food security will become more difficult to achieve as resources run out. Resources in danger of becoming depleted include oil, phosphorus, grain, fish, and water. The British scientist John Beddington predicted in 2009 that supplies of energy, food, and water will need to be increased by 50% to reach demand levels of 2030. According to the Food and Agriculture Organization (FAO), food supplies will need to be increased by 70% by 2050 to meet projected demands.

Africa

In Africa, if current trends of soil degradation and population growth continue, the continent might be able to feed just 25% of its population by 2025, according to UNU's Ghana-based Institute for Natural Resources in Africa.

Hunger and malnutrition kill nearly 6 million children a year, and more people are malnourished in sub-Saharan Africa this decade than in the 1990s, according to a report released by the Food and Agriculture Organization. In sub-Saharan Africa, the number of malnourished people grew to 203.5 million people in 2000–02 from 170.4 million 10 years earlier says *The State of Food Insecurity in the World* report. In 2001, 46.4% of people in sub-Saharan Africa were living in extreme poverty.

Dhaka street crowds. Bangladesh.

Asia

According to a 2004 article from the BBC, China, the world's most populous country, suffers from an "obesity surge". The article stated that, "Altogether, around 200 million people are thought to be overweight, 22.8% of the population, and 60 million (7.1%) obese". More recent data indicate China's grain production peaked in the mid-1990s, due to increased extraction of groundwater in the North China plain.

Other Countries

Nearly half of India's children are malnourished, according to recent government data. Japan may face a food crisis that could reduce daily diets to the austere meals of the 1950s, believes a senior government adviser.

Population as a Function of Food Availability

Thinkers from a wide range of academic fields and political backgrounds—including agricultural scientist David Pimentel, behavioral scientist Russell Hopfenberg, right-wing anthropologist Virginia Abernethy, ecologist Garrett Hardin, ecologist and anthropologist Peter Farb, journalist Richard Manning, environmental biologist Alan D. Thornhill, cultural critic and writer Daniel Quinn, and anarcho-primitivist John Zerzan,—propose that, like all other animal populations, human populations predictably grow and shrink according to their available food supply, growing during an abundance of food and shrinking in times of scarcity.

Proponents of this theory argue that every time food production is increased, the population grows. Most human populations throughout history validate this theory, as does the overall current global population. Populations of hunter-gatherers fluctuate in accordance with the amount of available food. The world human population began increasing after the Neolithic Revolution and its increased food supply. This was, subsequent to the Green Revolution, followed by even more severely accelerated population growth, which continues today. Often, wealthier countries send their surplus food resources to the aid of starving communities; however, proponents of this theory argue that this seemingly beneficial notion only results in further harm to those communities in the long run. Peter Farb, for example, has commented on the paradox that "intensification of production to feed an increased population leads to a still greater increase in population." Daniel Quinn has also focused on this phenomenon, which he calls the "Food Race" (comparable, in terms of both escalation and potential catastrophe, to the nuclear arms race).

Critics of this theory point out that, in the modern era, birth rates are lowest in the developed nations, which also have the highest access to food. In fact, some developed countries have both a diminishing population and an abundant food supply. The United Nations projects that the population of 51 countries or areas, including Germany, Italy, Japan, and most of the states of the former Soviet Union, is expected to be lower in 2050 than in 2005. This shows that, limited to the scope of the population living within a single given political boundary, particular human populations do not always grow to match the available food supply. However, the global population as a whole still grows in accordance with the total food supply and many of these wealthier countries are major *exporters* of food to poorer populations, so that, "it is through exports from food-rich to food-poor areas (Allaby, 1984; Pimentel et al., 1999) that the population growth in these food-poor areas is further fueled."

Regardless of criticisms against the theory that population is a function of food availability, the human population is, on the global scale, undeniably increasing, as is the net quantity of human food produced — a pattern that has been true for roughly 10,000 years, since the human development of agriculture. The fact that some affluent countries demonstrate negative population growth fails to discredit the theory as whole, since the world has become a globalized system with food moving across national borders from areas of abundance to areas of scarcity. Hopfenberg and Pimentel's findings support both this and Quinn's direct accusation that "First World farmers are fueling the Third World population explosion." Additionally, the hypothesis is not so simplistic as to be rejected by any single case study, as in Germany's recent population trends; clearly other factors are at work to limit the population in wealthier areas: contraceptive access, educational programs, cultural norms and, most influentially, differing economic realities from nation to nation.

As a Result of Water

Water deficits, which are already spurring heavy grain imports in numerous smaller countries, may soon do the same in larger countries, such as China or India, if technology is not used. The water tables are falling in scores of countries (including Northern China, the US, and India) owing to widespread overdrafting beyond sustainable yields. Other countries affected include Pakistan, Iran, and Mexico. This overdrafting is already leading to water scarcity and cutbacks in grain harvest. Even with the overpumping of its aquifers, China has developed a grain deficit. This effect has contributed in driving grain prices upward. Most of the 3 billion people projected to be added worldwide by mid-century will be born in countries already experiencing water shortages. One suggested solution is for population growth to be slowed quickly by investing heavily in female literacy and family planning services. Desalination is also considered a viable and effective solution to the problem of water shortages.

After China and India, there is a second tier of smaller countries with large water deficits – Algeria, Egypt, Iran, Mexico, and Pakistan. Four of these already import a large share of their grain. Only Pakistan remains self-sufficient. But with a population expanding by 4 million a year, it will also soon turn to the world market for grain.

Land

The World Resources Institute states that "Agricultural conversion to croplands and managed pastures has affected some 3.3 billion [hectares] – roughly 26 percent of the land area. All totaled, agriculture has displaced one-third of temperate and tropical forests and one-quarter of natural grasslands." Forty percent of the land area is under conversion and fragmented; less than one quarter, primarily in the Arctic and the deserts, remains intact. Usable land may become less useful through salinization, deforestation, desertification, erosion, and urban sprawl. Global warming may cause flooding of many of the most productive agricultural areas. The development of energy sources may also require large areas, for example, the building of hydroelectric dams. Thus, available useful land may become a limiting factor. By most estimates, at least half of cultivable land is already being farmed, and there are concerns that the remaining reserves are greatly overestimated.

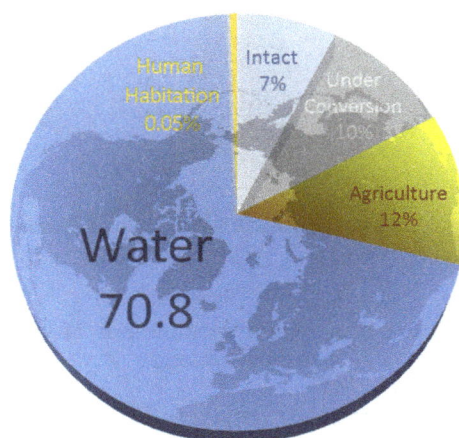

Percentages of the Earth's surface covered by water, dedicated to agriculture, under conversion, intact, and used for human habitation. While humans ourselves occupy only 0.05% of the Earth's total area, our effects are felt on over one-quarter of the land.

High crop yield vegetables like potatoes and lettuce use less space on inedible plant parts, like stalks, husks, vines, and inedible leaves. New varieties of selectively bred and hybrid plants have larger edible parts (fruit, vegetable, grain) and smaller inedible parts; however, many of these gain of agricultural technology are now historic, and new advances are more difficult to achieve. With new technologies, it is possible to grow crops on some marginal land under certain conditions. Aquaculture could theoretically increase available area. Hydroponics and food from bacteria and fungi, like quorn, may allow the growing of food without having to consider land quality, climate, or even available sunlight, although such a process may be very energy-intensive. Some argue that not all arable land will remain productive if used for agriculture because some marginal land can only be made to produce food by unsustainable practices like slash-and-burn agriculture. Even with the modern techniques of agriculture, the sustainability of production is in question.

Some countries, such as the United Arab Emirates and particularly the Emirate of Dubai have constructed large artificial islands, or have created large dam and dike systems, like the Netherlands, which reclaim land from the sea to increase their total land area. Some scientists have said that in the future, densely populated cities will use vertical farming to grow food inside skyscrapers. The notion that space is limited has been decried by skeptics, who point out that the Earth's population of roughly 6.8 billion people could comfortably be housed an area comparable in size to the state of Texas, in the United States (about 269,000 square miles or 696,706.80 square kilometres). However, the impact of humanity extends over a far greater area than that required simply for housing.

Fossil Fuels

Population optimists have been criticized for failing to take into account the depletion of the petroleum required for the production of fertilizers and fuel for transportation, as well as other fossil fuels. In his 1992 book *Earth in the Balance*, Al Gore wrote, "... it ought to be possible to establish a coordinated global program to accomplish the strategic goal of completely eliminating the internal combustion engine over, say, a twenty-five-year period..." Approximately half of the oil produced in the United States is refined into gasoline for use in internal combustion engines.

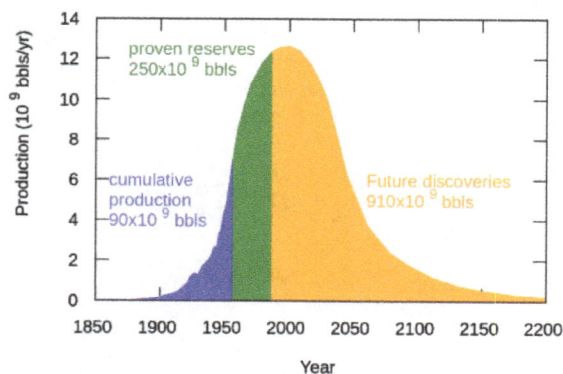

M. King Hubbert's prediction of world petroleum production rates. Modern agriculture is totally reliant on petroleum energy.

The report *Peaking of World Oil Production: Impacts, Mitigation, and Risk Management*, commonly referred to as the Hirsch report, was created by request for the US Department of Energy and published in February 2005. Some information was updated in 2007. It examined the time frame for the occurrence of peak oil, the necessary mitigating actions, and the likely impacts based on the timeliness of those actions. It concludes that world oil peaking is going to happen, and will likely be abrupt. Initiating a mitigation crash program 20 years before peaking appears to offer the possibility of avoiding a world liquid fuels shortfall for the forecast period.

Optimists counter that fossil fuels will be sufficient until the development and implementation of suitable replacement technologies—such as nuclear power or various sources of renewable energy—occurs. Methods of manufacturing fertilizers from garbage, sewage, and agricultural waste by using thermal depolymerization have been discovered.

Wealth and Poverty

Percentage living on less than $1 per day

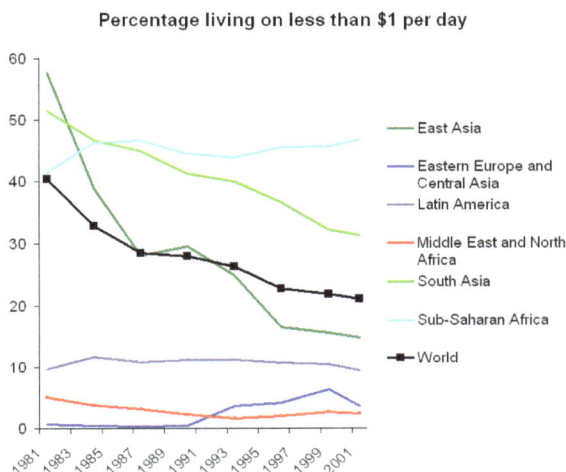

As the world's population has grown, the percentage of the world's population living on less than $1 per day (adjusted for inflation) has halved in 20 years. The graph shows the 1981–2001 period.

The United Nations indicates that about 850 million people are malnourished or starving, and 1.1 billion people do not have access to safe drinking water. Some argue that Earth may support 6 billion people, but only if many live in misery. The proportion of the world's population living on less than $1 per day has halved in 20 years, but these are inflation-unadjusted numbers and likely misleading. Since 1980, the global economy has grown by 380 percent, but the number of people living on less than $5 a day increased by more than 1.1 billion.

The UN Human Development Report of 1997 states: "During the last 15–20 years, more than 100 developing countries, and several Eastern European countries, have suffered from disastrous growth failures. The reductions in standard of living have been deeper and more long-lasting than what was seen in the industrialised countries during the depression in the 1930s. As a result, the income for more than one billion people has fallen below the level that was reached 10, 20 or 30 years ago". Similarly, although the proportion of "starving" people in sub-Saharan Africa has decreased, the absolute number of starving people has increased due to population growth. The percentage dropped from 38% in 1970 to 33% in 1996 and was expected to be 30% by 2010. But the region's population roughly doubled between 1970 and 1996. To keep the numbers of starving constant, the percentage would have dropped by more than half.

As of 2004, there were 108 countries in the world with more than five million people. All of these in which women have, on the average, more than 4 children in their lifetime, have a per capita GDP of less than $5000. Only in two countries with per capita GDP above ~$15,000 do women have, on the average, more than 2 children in their lifetime: these are Israel and Saudi Arabia, with average lifetime births per woman between 2 and 4.

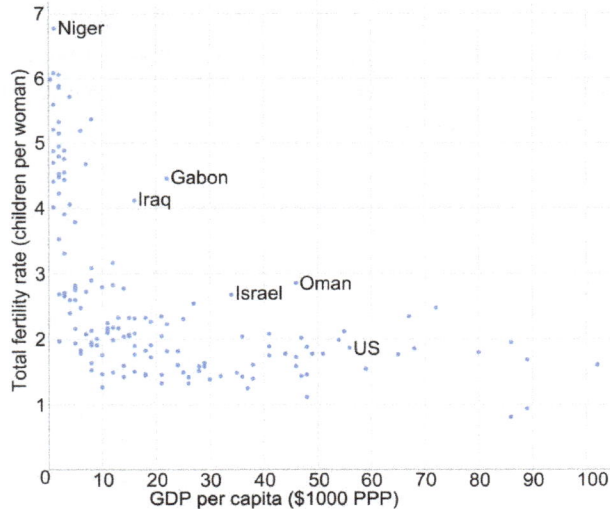

Graph of Total Fertility Rate vs. GDP (PPP) per capita of the corresponding country, 2015.

As their income increases, women are liberated and tend to have fewer "quantity kids", as in two in place of six.

The correlation does not imply cause and effect, and can be linked to the interplay of birth rates, death rates and economic development.

Poor living conditions can also cause a very bad effect on the population; diseases such as malaria and HIV/AIDS can also contribute to this. Lack of nutrients, poor sanitation and poor health institutions. Death rate and birth rate can also have a negative effect on the population.

Environment

Overpopulation has substantially adversely impacted the environment of Earth starting at least as early as the 20th century. According to the Global Footprint Network, "today humanity uses the equivalent of 1.5 planets to provide the resources we use and absorb our waste". There are also economic consequences of this environmental degradation in the form of ecosystem services attrition. Beyond the scientifically verifiable harm to the environment, some assert the moral right of other species to simply exist rather than become extinct. Environmental author Jeremy Rifkin has said that "our burgeoning population and urban way of life have been purchased at the expense of vast ecosystems and habitats. ... It's no accident that as we celebrate the urbanization of the world, we are quickly approaching another historic watershed: the disappearance of the wild."

Says Peter Raven, former President of the American Association for the Advancement of Science (AAAS) in their seminal work AAAS Atlas of Population & Environment, "Where do we stand in our efforts to achieve a sustainable world? Clearly, the past half century has been a traumatic one, as the collective impact of human numbers, affluence (consumption per individual) and our choices of technology continue to exploit rapidly an increasing proportion of the world's resources at an unsustainable rate. ... During a remarkably short period of time, we have lost a quarter of the world's topsoil and a fifth of its agricultural land, altered the composition of the atmosphere profoundly, and destroyed a major proportion of our forests and other natural habitats without replacing them. Worst of all, we have driven the rate of biological extinction, the permanent loss of species, up several hundred times beyond its historical levels, and are threatened with the loss of a majority of all species by the end of the 21st century."

Traffic congestion in Ho Chi Minh City, Vietnam

Further, even in countries which have both large population growth and major ecological problems, it is not necessarily true that curbing the population growth will make a major contribution towards resolving all environmental problems. However, as developing countries with high populations become more industrialized, pollution and consumption will invariably increase.

The Worldwatch Institute said in 2006 that the booming economies of China and India are "planetary powers that are shaping the global biosphere". The report states:

The world's ecological capacity is simply insufficient to satisfy the ambitions of China, India, Japan, Europe and the United States as well as the aspirations of the rest of the world in a sustainable way.

It is said that if China and India were to consume as much resources per capita as the United States, in 2030 they would each require a full planet Earth to meet their needs. In the long term these effects can lead to increased conflict over dwindling resources and in the worst case a Malthusian catastrophe.

Many studies link population growth with emissions and the effect of climate change.

Warfare and Conflict

" excessive growth may reduce output per worker, repress levels of living for the masses and engender strife. Confucius (551 – 479 BC) "

" Overpopulation in various countries has become a serious threat to the health of people and a grave obstacle to any attempt to organize peace on this planet. Albert Einstein – physicist 1879 – 1955 "

It has been suggested that overpopulation leads to increased levels of tensions both between and within countries. Modern usage of the term "lebensraum" supports the idea that overpopulation may promote warfare through fear of resource scarcity and increasing numbers of youth lacking the opportunity to engage in peaceful employment (the youth bulge theory).

Criticism of this Hypothesis

The hypothesis that population pressure causes increased warfare has been recently criticized on the empirical grounds. Both studies focusing on specific historical societies and analyses of cross-cultural data have failed to find positive correlation between population density and incidence of warfare. Andrey Korotayev, in collaboration with Peter Turchin, has shown that such negative results do not falsify the population-warfare hypothesis.

Population and warfare are dynamical variables, and if their interaction causes sustained oscillations, then we do not in general expect to find strong correlation between the two variables measured at the same time (that is, unlagged). Korotayev and Turchin have explored mathematically what the dynamical patterns of interaction between population and warfare (focusing on internal warfare) might be in both stateless and state societies. Next, they have tested the model predictions in several empirical case studies: early modern England, Han and Tang China, and the Roman Empire. Their empirical results have supported the population-warfare theory: that there is a tendency for population numbers and internal warfare intensity to oscillate with the same period but shifted in phase (with warfare peaks following population peaks).

Furthermore, they have demonstrated that in the agrarian societies the rates of change of the two variables behave precisely as predicted by the theory: population rate of change is negatively affected by warfare intensity, while warfare rate of change is positively affected by population density.

Mitigation Measures

There are several mitigation measures that have been or can be applied to reduce overpopulation. All of these mitigations are ways to implement social norms. Overpopulation is an issue that threatens the state of the environment in the above-mentioned ways and therefore societies must make a change in order to reverse some of the environmental effects brought on by current social norms. In societies like China, the government has put policies in place that regulate the number of children allowed to a couple. Other societies have already begun to implement social marketing strategies in order to educate the public on overpopulation effects. "The intervention can be widespread and done at a low cost. A variety of print materials (flyers, brochures, fact sheets, stickers) needs to be produced and distributed throughout the communities such as at local places of worship, sporting events, local food markets, schools and at car parks (taxis / bus stands)."

Such prompts work to introduce the problem so that social norms are easier to implement. Certain government policies are making it easier and more socially acceptable to use contraception and abortion methods. An example of a country whose laws and norms are hindering the global effort to slow population growth is Afghanistan. "The approval by Afghan President Hamid Karzai of the Shia Personal Status Law in March 2009 effectively destroyed Shia women's rights and freedoms in Afghanistan. Under this law, women have no right to deny their husbands sex unless they are ill, and can be denied food if they do."

Education and Empowerment

One option is to focus on education about overpopulation, family planning, and birth control methods, and to make birth-control devices like male/female condoms, pills and intrauterine devices easily available. Worldwide, nearly 40% of pregnancies are unintended (some 80 million unintended pregnancies each year). An estimated 350 million women in the poorest countries of the world either did not want their last child, do not want another child or want to space their pregnancies, but they lack access to information, affordable means and services to determine the size and spacing of their families. In the United States, in 2001, almost half of pregnancies were unintended. In the developing world, some 514,000 women die annually of complications from pregnancy and abortion, with 86% of these deaths occurring in the sub-Saharan Africa region and South Asia. Additionally, 8 million infants die, many because of malnutrition or preventable diseases, especially from lack of access to clean drinking water.

Women's rights and their reproductive rights in particular are issues regarded to have vital importance in the debate.

"The only ray of hope I can see – and it's not much – is that wherever women are put in control of their lives, both politically and socially; where medical facilities allow them to deal with birth control and where their husbands allow them to make those decisions, birth rate falls. Women don't want to have 12 kids of whom nine will die." — David Attenborough

Egypt announced a program to reduce its overpopulation by family planning education and putting women in the workforce. It was announced in June 2008 by the Minister of Health and Population Hatem el-Gabali. The government has set aside 480 million Egyptian pounds (about $90 million US) for the program.

The business magnate Ted Turner proposed a "voluntary, non-imposed" one-child family scenario. A "pledge two or fewer" campaign is run by Population Matters (a UK population concern organisation), in which people are encouraged to limit themselves to small family size.

Birth Regulations

Overpopulation is related to the issue of birth control; some nations, like the People's Republic of China, use strict measures to reduce birth rates. Religious and ideological opposition to birth control has been cited as a factor contributing to overpopulation and poverty.

Sanjay Gandhi, son of late Prime Minister of India Indira Gandhi, implemented a forced sterilization programme between 1975 and 1977. Officially, men with two children or more had to submit to sterilization, but there was a greater focus on sterilizing women than sterilizing men. Some unmarried young men, political opponents and ignorant men were also believed to have been sterilized. This program is still remembered and criticized in India, and is blamed for creating a public aversion to family planning, which hampered government programmes for decades.

Urban designer Michael E. Arth has proposed a "choice-based, marketable birth license plan" he calls "birth credits". Birth credits would allow any woman to have as many children as she wants, as long as she buys a license for any children beyond an average allotment that would result in zero population growth. If that allotment was determined to be one child, for example, then the first child would be free, and the market would determine what the license fee for each additional child would cost. Extra credits would expire after a certain time, so these credits could not be hoarded by speculators. The actual cost of the credits would only be a fraction of the actual cost of having and raising a child, so the credits would serve more as a wake-up call to women who might otherwise produce children without seriously considering the long term consequences to themselves or society.

Population Growth in More- and Less-Developed Countries, 2002.

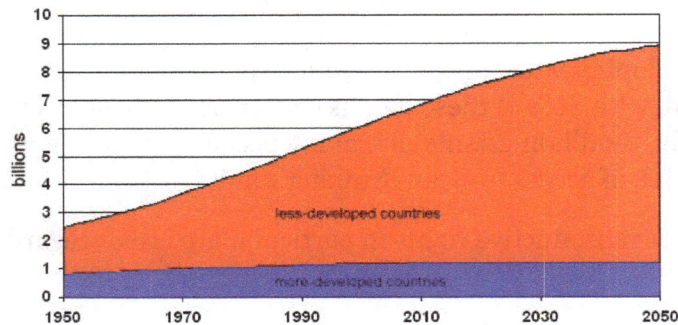

Source: United Nations, World Population Prospects.

Another choice-based approach, similar to Arth's birth credits, is financial compensation or other benefits (free goods and/or services) by the state (or state-owned companies) offered to people who voluntarily undergo sterilization. Such compensation has been offered in the past by the government of India.

In 2014 the United Nations estimated there is an 80% likelihood that the world's population will be between 9.6 billion and 12.3 billion by 2100. Most of the world's expected population increase will be in Africa and southern Asia. Africa's population is expected to rise from the current one billion to three or four billion by 2100, and Asia could add another billion in the same period. Because the median age of Africans is so low (e.g. Uganda = 15 years old) birth credits would have to limit fertility to one child per two women to reach the levels of developed countries immediately. For countries with a wide base in their population pyramid it will take a generation for the people who are of child bearing age to have their families. An example of demographic momentum is China, which added perhaps 400,000 more people after its one-child policy was enacted. Arth has suggested that the focus should be on the developed countries and that some combination of birth credits and additional compensation supplied by the developed countries could rapidly lead to zero population growth while also quickly raising the standard of living in developing countries.

Infectious Agents

There are also natural diseases, such as chlamydia, gonorrhea and even some diseases that produce no physical discomfort at all (such as some specific adeno-associated viruses) that may also cause sterilization of males and/or females in the population. Whereas the natural occurrence of these diseases is not sufficient to significantly reduce the population problem, large-scale production of these diseases and human-induced release of such diseases could quite well be used as a mitigation measure.

Extraterrestrial Settlement

Various scientists and science fiction authors have contemplated that overpopulation on Earth may be remedied in the future by the use of extraterrestrial settlements. In the 1970s, Gerard K.

O'Neill suggested building space habitats that could support 30,000 times the carrying capacity of Earth using just the asteroid belt, and that the Solar System as a whole could sustain current population growth rates for a thousand years. Marshall Savage (1992, 1994) has projected a human population of five quintillion (5×10^{18}) throughout the Solar System by 3000, with the majority in the asteroid belt. Freeman Dyson (1999) favours the Kuiper belt as the future home of humanity, suggesting this could happen within a few centuries. In *Mining the Sky*, John S. Lewis suggests that the resources of the solar system could support 10 quadrillion (10^{16}) people. In an interview, Stephen Hawking claimed that overpopulation is a threat to human existence and "our only chance of long-term survival is not to remain inward looking on planet Earth but to spread out into space."

K. Eric Drexler, famous inventor of the futuristic concept of molecular nanotechnology, has suggested in *Engines of Creation* that colonizing space will mean breaking the Malthusian limits to growth for the human species.

It may be possible for other parts of the Solar System to be inhabited by humanity at some point in the future. Geoffrey Landis of NASA's Glenn Research Center in particular has pointed out that "[at] cloud-top level, Venus is the paradise planet", as one could construct aerostat habitats and floating cities there easily, based on the concept that breathable air is a lifting gas in the dense Venusian atmosphere. Venus would, like also Saturn, Uranus, and Neptune, in the upper layers of their atmospheres, even afford a gravitation almost exactly as strong as that on Earth.

Many science fiction authors, including Carl Sagan, Arthur C. Clarke, and Isaac Asimov, have argued that shipping any excess population into space is not a viable solution to human overpopulation. According to Clarke, "the population battle must be fought or won here on Earth". The problem for these authors is not the lack of resources in space (as shown in books such as *Mining the Sky*), but the physical impracticality of shipping vast numbers of people into space to "solve" overpopulation on Earth. However, Gerard K. O'Neill's calculations show that Earth could offload all new population growth with a launch services industry about the same size as the current airline industry.

The StarTram concept, by James R. Powell (the co-inventor of maglev transport) and others, envisions a capability to send up to 4 million people a decade to space per facility. A hypothetical extraterrestrial colony could potentially grow by reproduction only (i.e., without any immigration), with all of the inhabitants being the direct descendants of the original colonists.

Urbanization

Despite the increase in population density within cities (and the emergence of megacities), UN Habitat states in its reports that urbanization may be the best compromise in the face of global population growth. Cities concentrate human activity within limited areas, limiting the breadth of environmental damage. But this mitigating influence can only be achieved if urban planning is significantly improved and city services are properly maintained.

References

- Taylor, Leslie (2004). The Healing Power of Rainforest Herbs: A Guide to Understanding and Using Herbal Medicinals. Square One. ISBN 9780757001444.

- F. Terry Norris, "Where Did the Villages Go? Steamboats, Deforestation, and Archaeological Loss in the Mississippi Valley", in Common Fields: an environmental history of St. Louis, Andrew Hurley, ed., St. Louis, MO: Missouri Historical Society Press, 1997, ISBN 978-1-883982-15-7 pp. 73–89

- Ron Nielsen, The Little Green Handbook: Seven Trends Shaping the Future of Our Planet, Picador, New York (2006) ISBN 978-0-312-42581-4

- John F. Mongillo; Linda Zierdt-Warshaw (2000). Linda Zierdt-Warshaw, ed. Encyclopedia of environmental science. University of Rochester Press. p. 104. ISBN 978-1-57356-147-1.

- Ryerson, W. F. (2010). "Population, The Multiplier of Everything Else". In McKibben, D. The Post Carbon Reader: Managing the 21st Century Sustainability Crisis. Watershed Media. ISBN 978-0-9709500-6-2

- Leakey, Richard and Roger Lewin, 1996, The Sixth Extinction : Patterns of Life and the Future of Humankind, Anchor, ISBN 0-385-46809-1

- Ron Nielsen, The Little Green Handbook: Seven Trends Shaping the Future of Our Planet, Picador, New York (2006) ISBN 978-0-312-42581-4

- Worldwatch, The. Outgrowing the Earth: The Food Security Challenge in an Age of Falling Water Tables and Rising Temperatures: Books: Lester R. Brown. Amazon.com. ISBN 0393060705.

- Daily, Gretchen C. and Ellison, Katherine (2003) The New Economy of Nature: The Quest to Make Conservation Profitable, Island Press ISBN 1559631546

- Champion, Tony (2005). "Chapter 4: Demographic transformations". In Daniels, Peter; Bradshaw, Michael; Shaw, Denis; Sidaway, James. An Introduction to Human Geography Issues for the 21st Century Second edition. Pearson Education. pp. 88–111. ISBN 0-131-21766-6.

- Korotayev A.V., Khaltourina D.A. Introduction to Social Macrodynamics: Secular Cycles and Millennial Trends in Africa. Moscow: URSS, 2006. ISBN 5-484-00560-4;

- "Stolen Goods: The EU's complicity in illegal tropical deforestation" (PDF). Forests and the European Union Resource Network. March 17, 2015. Retrieved March 31, 2015.

- "How much has the Global Temperature Risen in the Last 100 Years?". National Center for Atmospheric Research. University Corporation for Atmospheric Research. Retrieved 20 October 2014.

- "Study: Loss Of Genetic Diversity Threatens Species Diversity". Environmental News Network. 26 September 2007. Retrieved 27 October 2014.

- David Pimentel; Mario Giampietro (21 November 1994). FOOD, LAND, POPULATION and the U.S. ECONOMY (Report). Washington, D.C.: Carrying Capacity Network. Retrieved 2014-09-07.

Conservation Measures

To mitigate the damage to Earth's biodiversity, local governments have taken counter measures like the opening of animal sanctuaries, national parks, biodiversity banking, biodiversity offsetting and mitigation banking. These measures have begun a movement in the right direction and in this chapter readers are informed about how these steps have been carried out.

National Park

A national park is a park in use for conservation purposes. Often it is a reserve of natural, semi-natural, or developed land that a sovereign state declares or owns. Although individual nations designate their own national parks differently, there is a common idea: the conservation of 'wild nature' for posterity and as a symbol of national pride. An international organization, the International Union for Conservation of Nature (IUCN), and its World Commission on Protected Areas, has defined "National Park" as its *Category II* type of protected areas.

An elephant safari through the Jaldapara National Park in West Bengal, India

While this type of national park had been proposed previously, the United States established the first "public park or pleasuring-ground for the benefit and enjoyment of the people", Yellowstone National Park, in 1872. Although Yellowstone was not officially termed a "national park" in its establishing law, it was always termed such in practice and is widely held to be the first and oldest national park in the world. The first area to use "national park" in its creation legislation was the US's Mackinac Island, in 1875. Australia's Royal National Park, established in 1879, was the world's third official national park. In 1895 ownership of Mackinac Island was transferred to the State of

Michigan as a state park and national park status was consequently lost. As a result, Australia's Royal National Park is by some considerations the second oldest national park now in existence.

The Teide National Park in Tenerife, Spain

The largest national park in the world meeting the IUCN definition is the Northeast Greenland National Park, which was established in 1974. According to the IUCN, 6,555 national parks worldwide met its criteria in 2006. IUCN is still discussing the parameters of defining a national park.

National parks are almost always open to visitors. Most national parks provide outdoor recreation and camping opportunities as well as classes designed to educate the public on the importance of conservation and the natural wonders of the land in which the national park is located.

Definitions

Manuel Antonio National Park in Costa Rica was listed by *Forbes* as one of the world's 12 most beautiful national parks.

A salt marsh in Schiermonnikoog National Park, Netherlands

In 1969, the IUCN declared a national park to be a relatively large area with the following defining characteristics:

- One or several ecosystems not materially altered by human exploitation and occupation, where plant and animal species, geomorphological sites and habitats are of special scientific, educational, and recreational interest or which contain a natural landscape of great beauty;

- Highest competent authority of the country has taken steps to prevent or eliminate exploitation or occupation as soon as possible in the whole area and to effectively enforce the respect of ecological, geomorphological, or aesthetic features which have led to its establishment; and

- Visitors are allowed to enter, under special conditions, for inspirational, educative, cultural, and recreative purposes.

In 1971, these criteria were further expanded upon leading to more clear and defined benchmarks to evaluate a national park. These include:

- Minimum size of 1,000 hectares within zones in which protection of nature takes precedence

- Statutory legal protection

- Budget and staff sufficient to provide sufficient effective protection

- Prohibition of exploitation of natural resources (including the development of dams) qualified by such activities as sport, fishing, the need for management, facilities, etc.

While the term national park is now defined by the IUCN, many protected areas in many countries are called national park even when they correspond to other categories of the IUCN Protected Area Management Definition, for example:

- Swiss National Park, Switzerland: IUCN Ia - Strict Nature Reserve

- Everglades National Park, United States: IUCN Ib - Wilderness Area

- Białowieża National Park, Poland: IUCN II National Park

- Victoria Falls National Park, Zimbabwe: IUCN III - National Monument

- Vitosha National Park, Bulgaria: IUCN IV - Habitat Management Area

- New Forest National Park, United Kingdom: IUCN V - Protected Landscape

- Etniko Ygrotopiko Parko Delta Evrou, Greece: IUCN VI - Managed Resource Protected Area

While national parks are generally understood to be administered by national governments (hence the name), in Australia national parks are run by state governments and predate the Federation of Australia; similarly, national parks in the Netherlands are administered by the provinces. In many countries, including Indonesia, the Netherlands, and the United Kingdom, national parks do not adhere to the IUCN definition, while some areas which adhere to the IUCN definition are not designated as national parks.

History

In 1810, the English poet William Wordsworth described the Lake District as a sort of national property, in which every man has a right and interest who has an eye to perceive and a heart to enjoy.

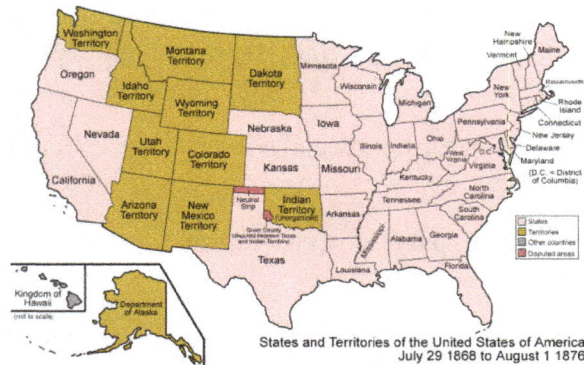

States and Territories of the United States of America
July 29 1868 to August 1 1876

The United States in 1872. When Yellowstone was established, Wyoming, Montana and Idaho were territories, not states. For this reason, the federal government had to assume responsibility for the land, hence the creation of the *national* park.

The painter George Catlin, in his travels through the American West, wrote during the 1830s that the Native Americans in the United States might be preserved (by some great protecting policy of government) ...in a *magnificent park* ...A *nation's Park*, containing man and beast, in all the wild and freshness of their nature's beauty!

The first effort by the Federal government to set aside such protected lands was on April 20, 1832, when President Andrew Jackson signed legislation that the 22nd United States Congress had enacted to set aside four sections of land around what is now Hot Springs, Arkansas, to protect the natural, thermal springs and adjoining mountainsides for the future disposal of the U.S. government. It was known as Hot Springs Reservation, but no legal authority was established. Federal control of the area was not clearly established until 1877.

Yosemite Valley, Yosemite National Park, in California

John Muir is today referred to as the "Father of the National Parks" due to his work in Yosemite. He published two influential articles in The Century Magazine, which formed the base for the subsequent legislation.

President Abraham Lincoln signed an Act of Congress on July 1, 1864, ceding the Yosemite Valley and the Mariposa Grove of Giant Sequoias (later becoming Yosemite National Park) to the state of California. According to this bill, private ownership of the land in this area was no longer possible. The state of California was designated to manage the park for "public use, resort, and recreation". Leases were permitted for up to ten years and the proceeds were to be used for conservation and improvement. A public discussion followed this first legislation of its kind and there was a heated debate over whether the government had the right to create parks. The perceived mismanagement of Yosemite by the Californian state was the reason why Yellowstone at its establishment six years later was put under national control.

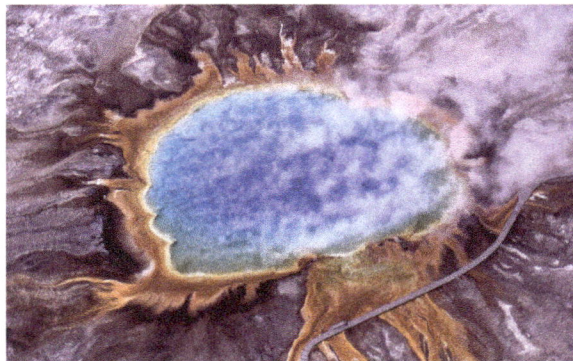

Grand Prismatic Spring in Yellowstone National Park, Wyoming; Yellowstone was the first national park in the world.

Los Cardones National Park in Salta province, Argentina

In 1872, Yellowstone National Park was established as the United States' first national park, being also the world's first national park. In some European countries, however, national protection and nature reserves already existed, such as Drachenfels (Germany, 1822) and a part of Forest of Fontainebleau (France, 1861).

Yellowstone was part of a federally governed territory. With no state government that could assume stewardship of the land so the federal government took on direct responsibility for the park, the official first national park of the United States. The combined effort and interest of conservationists, politicians and the Northern Pacific Railroad ensured the passage of enabling legislation by the United States Congress to create Yellowstone National Park. Theodore Roosevelt, already an active campaigner and so influential, as good stump speakers were highly necessary in the pre-telecommunications era, was highly influential in convincing fellow Republicans and big business to back the bill.

American Pulitzer Prize-winning author Wallace Stegner wrote:

National parks are the best idea we ever had. Absolutely American, absolutely democratic, they reflect us at our best rather than our worst.

In his book *Dispossessing the Wilderness: Indian Removal and the Making of the National Parks*, Mark David Spence made the point that in order to create these uninhabited spaces, the United States first had to disposess the Indians who were living in them.

Even with the creation of Yellowstone, Yosemite, and nearly 37 other national parks and monuments, another 44 years passed before an agency was created in the United States to administer these units in a comprehensive way – the U.S. National Park Service (NPS). The 64th United States Congress passed the National Park Service Organic Act, which President Woodrow Wilson signed into law on August 25, 1916. Of the 412 sites managed by the National Park Service of the United States, only 59 carry the designation of National Park.

Following the idea established in Yellowstone, there soon followed parks in other nations. In Australia, the Royal National Park was established just south of Sydney on April 26, 1879, becoming the world's second official national park (actually the 3rd: Mackinac National Park in Michigan was created in 1875 as a national park but was later transferred to the state's authority in 1895, thus losing its official "national park" status). Rocky Mountain National Park became Canada's first national park in 1885. Argentina became the third country in the Americas to create a national park system, with the creation of the Nahuel Huapi National Park in 1934, through the initiative of Francisco Moreno. New Zealand established Tongariro National Park in 1887. In Europe, the first national parks were a set of nine parks in Sweden in 1909, followed by the Swiss National Park in 1914. Europe has some 359 national parks as of 2010. Africa's first national park was established in 1925 when Albert I of Belgium designated an area of what is now Democratic Republic of Congo centred on the Virunga Mountains as the Albert National Park (since renamed Virunga National Park). In 1973, Mount Kilimanjaro was classified as a National Park and was opened to public access in 1977. In 1926, the government of South Africa designated Kruger National Park as the nation's first national park, although it was an expansion of the earlier Sabie Game Reserve established in 1898 by President Paul Kruger of the old South African Republic, after whom the park was named. After World War II, national parks were founded all over the world. The Vanoise National Park in the Alps was the first French national park, created in 1963 after public mobilization against a touristic project.

The world's first national park service was established May 19, 1911, in Canada. The Dominion Forest Reserves and Parks Act placed the dominion parks under the administration of the Dominion Park Branch (now Parks Canada). The branch was established to "protect sites of natural wonder" to provide a recreational experience, centered on the idea of the natural world providing rest and spiritual renewal from the urban setting. Canada now has the largest protected area in the world with 377,000 km² of national park space. In 1989, the Qomolangma National Nature Preserve (QNNP) was created to protect 3.381 million hectares on the north slope of Mount Everest in the

Tibet Autonomous Region of China. This national park is the first major global park to have no separate warden and protection staff—all of its management being done through existing local authorities, allowing a lower cost basis and a larger geographical coverage (in 1989 when created, it was the largest protected area in Asia). It includes four of the six highest mountains Everest, Lhotse, Makalu, and Cho Oyu. The QNNP is contiguous to four Nepali national parks, creating a transborder conservation area equal in size to Switzerland.

- National parks

Lower Consolation Lake in Banff National Park, Alberta, Canada

Hadrian's Wall crosses Northumberland National Park in England

Serra dos Órgãos National Park in Rio de Janeiro state, Brazil. This national park is one of the few natural habitats of species of *Schlumbergera*

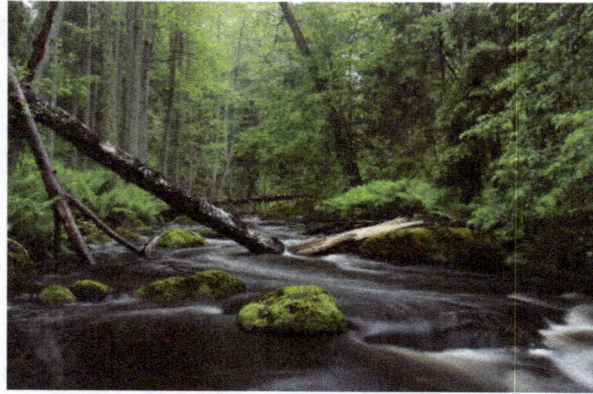

Lahemaa National Park in Estonia was the first area to be designated a national park in the former Soviet Union

Economic Ramifications

Countries with a large nature-based tourism industry, such as Costa Rica, often experience a huge economic effect on park management as well as the economy of the country as a whole.

Tourism

Tourism to national parks has increased considerably over time. In Costa Rica for example, a mega-diverse country, tourism to parks has increased by 400% from 1985 to 1999. The term *national park* is perceived as a brand name that is associated with nature-based tourism and it symbolizes "high quality natural environment and well-design tourism infrastructure".

Staff

The duties of a park ranger are to supervise, manage, and/or perform work in the conservation and use of Federal park resources. This involves functions such as park conservation; natural, historical, and cultural resource management; and the development and operation of interpretive and recreational programs for the benefit of the visiting public. Park rangers also have fire fighting responsibilities and execute search and rescue missions. Activities also include heritage interpretation to disseminate information to visitors of general, historical, or scientific information. Management of resources such as wildlife, lakeshores, seashores, forests, historic buildings, battlefields, archeological properties, and recreation areas are also part of the job of a park ranger. Since the establishment of the National Park Service in the US in 1916, the role of the park ranger has shifted from merely being a custodian of natural resources to include several activities that are associated with law enforcement. They control traffic and investigate violations, complaints, trespass/encroachment, and accidents.

Animal Sanctuary

An animal sanctuary is a facility where animals are brought to live and be protected for the rest of their lives. Unlike animal shelters, sanctuaries do not seek to place animals with individuals or groups, instead maintaining each animal until his or her natural death. In some cases, an establishment may have characteristics of both a sanctuary and a shelter; for instance, some animals may be in residence temporarily until a good home is found and others may be permanent residents. The mission of sanctuaries is generally to be safe havens, where the animals receive the best care that the sanctuaries can provide. Animals are not bought, sold, or traded, nor are they used for animal testing. The resident animals are given the opportunity to behave as natural as possible in a protective environment.

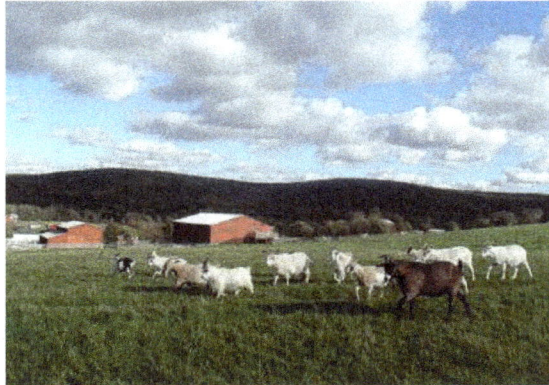

Farm Sanctuary's shelter in upstate New York provides a home to hundreds of rescued goats, sheep, cows, pigs, chickens, turkeys, and other farm animals.

What distinguishes a sanctuary from other institutions is the philosophy that the residents come first. In a sanctuary, every action is scrutinized for any trace of human benefit at the expense of non-human residents. Sanctuaries act on behalf of the animals, and the caregivers work under the notion that all animals in the sanctuary, human and non-human, are of equal importance.

A sanctuary is not open to the public in the sense of a zoo; that is, the public is not allowed unescorted access to any part of the facility. A sanctuary tries not to allow any activity that would place the animals in an unduly stressful situation.

One of the most important missions of sanctuaries, beyond caring for the animals, is educating the public. The ultimate goal of many sanctuaries is to change the way that humans think of, and treat, non-human animals.

Biodiversity Banking

Biodiversity banking, also known as biodiversity trading or conservation banking, biodiversity mitigation banks, compensatory habitat, set-asides, biodiversity offsets, are conservation activities that compensate for the loss of biodiversity with the goal of biodiversity reduction through a framework which allows biodiversity to be reliably measured, and market based solutions applied to improving biodiversity. Biodiversity banking provides a means to place a monetary value on ecosystem services. Typically this involves land protection, restoration, an/or enhancement. Biodiversity banking is often applied so that there is no "net loss of a particular biodiversity feature." According to the International Union for Conservation of Nature, by 2004, interest in voluntary biodiversity offsets was growing in the United States, Brazil, Australia, Canada and the EU.

Experience suggested that industry, governments, local communities and conservation groups all benefit from biodiversity offsets or biodiversity banking.

Terminology

According to the IUCN, "[b]iodiversity offsets are designed to compensate for residual environmental damage caused by development after avoidance, minimization, and mitigation of environmental impacts have been considered and implemented. The goal of offsets is to compensate for the loss of biodiversity at one location with conservation gains elsewhere."

In Practice

In practice, biodiversity banks rely on existing governmental laws, which forbid companies or individuals of buying up land in an area that houses, say a critically endangered species. An exception is also in place in this governmental law which allows companies to buy up the land nonetheless, if they also buy a certain amount of compensation credits with a certified biodiversity bank. These credits, which represent a significant extra cost to the company, are then used to provide revenue for the biodiversity bank, but the revenue derived thereof is also used by the bank to buy up conservation area elsewhere for the endangered species.

United States

In the United States a "mitigation banking" process applies to impacts on wetlands. It requires that developers firstly avoid harm to wetlands, but if harm is considered unavoidable, then wetland habitat of similar function and values must be "protected, enhanced or restored" to compensate for those that will be damaged. The process comes under the US Clean Water Act 1972 the US Army Corps of Engineers regulations and the commitment to "no net loss" of wetlands habitat.

Since about 2000 the term "species banking", sometimes called "conservation banking", has applied to impacts on species of special concern, typically those that are listed by state and federal agencies under the U.S. Endangered Species Act or its state-based equivalent. Similar to wetlands banks, conservation banks are designed as compensation for impacts to listed species or their habitat, ensuring a similar no net loss policy for these biodiversity resources.

Compensation for impacts to a stream riparian zone may also be required in relation to the linear distance of lost stream functions resulting from stream bank structures (e.g., concrete or rip rap), sedimentation, channelization, dredging or similar activities.

Australia

Two biodiversity banking schemes operating in Australia are the New South Wales BioBanking scheme, which commenced in 2008, and the Victorian Native Vegetation Management Framework scheme. Both schemes apply particularly to developers, where biodiversity values will be reduced through land clearing and building development. The framework requires developers to source biodiversity credits through a market mechanism to offset biodiversity loss.

Listed species, critical habitat, wetlands and stream habitat are all components of biodiversity ecosystem services. Taken collectively, they may be referred to as "biodiversity banks".

Canada

In Alberta, Canada, Alberta Biodiversity Monitoring Institute (ABMI) researchers use the oil sands industry of Alberta as a case study in their paper in which they evaluated the commonly used and costly ecological equivalency-based biodiversity offset in terms of economic and ecological performance with more flexible alternative offset systems. They used ABMI's "empirically derived index of biodiversity intactness to link offsets with losses incurred by development." They evaluated ecologically equivalent areas in regards to vegetation types and regional conservation priorities such as the recovery of the boreal woodland caribou and the Dry Mixedwood natural subregion in the oil sands region. They found that flexible alternative systems like the priority-focused offsetting networks, cost 2-17 times less than the ecological equivalency-based biodiversity offset vegetation cost 2–17 times more than priority-focused networks.

Biodiversity Offsetting

Biodiversity offsetting is a system used predominantly by planning authorities and developers to fully compensate for biodiversity impacts associated with economic development, through the planning process. In some circumstances, biodiversity offsets are designed to result in an overall biodiversity gain. Offsetting is generally considered the final stage in a mitigation hierarchy, whereby predicted biodiversity impacts must first be avoided, minimised and reversed by developers, before any remaining impacts are offset. The mitigation hierarchy serves to meet the environmental policy principle of "No Net Loss" of biodiversity alongside development.

Bringing forward farmland sites to receive biodiversity offset credits will create the investment needed to improve biodiversity across large areas.

Individuals or companies involved in arranging biodiversity offsets will use quantitative measures to determine the amount, type and quality of habitat that is likely to be affected by a proposed project. Then, they will establish a new location or locations (often called receptor sites) where it would be possible to re-create the same amount, type and quality of habitat. The aim of biodiversity offsets is not simply to provide financial compensation for the biodiversity losses associated with development, although developers might pay financial compensation in some cases if it can be demonstrated exactly what the physical biodiversity gains achieved by that compensation will be. The type of environmental compensation provided by biodiversity offsetting is different from similar systems in that it must show both measurable and long-term biodiversity improvements, that can be demonstrated to counteract losses.

Relevant Conservation Activities

Biodiversity offset projects can involve various management activities that can be demonstrated to deliver gains in biodiversity. These activities very often include active habitat restoration or creation projects (e.g. new wetland creation, grassland restoration). However, also viable are so-called "averted loss" biodiversity offsets, in which measures are taken to prevent ecological degradation from occurring where it almost certainly would have happened otherwise. Averted loss offsets might involve the creation of new protected areas (to conserve fauna species that would otherwise have disappeared), the removal of invasive species from areas of habitat (which otherwise would have reduced or displaced populations of native species), or positive measures to reduce extensive natural resource use (e.g. the offer of alternative livelihood creation to prevent activities leading to deforestation).

Any activities that do not result in a positive and measurable gain for biodiversity would not generally be counted as part of a biodiversity offset. For instance, if a developer funds ecological conservation research in a region that they are impacting through a project, would not count as an offset (unless it could be shown quantitatively how specific fauna and flora would benefit). instead, this would be a more general form of compensation. Note that biodiversity offsets can be considered a very specific, robust and transparent category of ecological compensation.

Receptor Sites

Under many offset systems, receptor sites are areas of land put forward by companies or individuals looking to receive payment in return for creating (or restoring) biodiversity habitats on their property. The biodiversity restoration projects are financed by compensation from developers looking to offset their biodiversity impact. The resulting change in biodiversity levels at the new receptor sites should be equal to, or greater than, the losses at the original 'impact site'; in order to achieve no net loss – and preferably gain – of overall biodiversity. Such systems often rely on the buying (by developers) and selling (by landowners) of conservation credits.

However, characteristics of receptor sites can vary across different jurisdictions. In some countries, for instance, land is primarily state-owned, and so it is the government that owns and manages biodiversity offset projects. For biodiversity offsets in marine environments, receptor sites might be subject to multiple management organisations and not necessarily owned by anyone. Controversially, some biodiversity offsets use existing protected areas as receptor sites (i.e. improving the effectiveness of areas that are already managed for biodiversity conservation).

Requirement to Offset Biodiversity

Biodiversity offsets are required by law in many jurisdictions (Madsen et al., 2011).

Countries including the US, Australia, New Zealand, UK and parts of Europe use biodiversity offsetting as an optional or mandatory (depending on the country) biodiversity conservation management tool within their planning systems.

Biodiversity offsetting is also being considered by some Latin American countries (Colombia, Peru, Ecuador and Chile) and by South Africa.

Another key driver of biodiversity offset projects globally are the Performance Standards required by the International Finance Corporation (IFC). For any projects which the IFC or any of the Equator Banks 'Equator Banks' finance, under Performance Standard 6, developers must deliver No Net Loss (or in some cases, a Net Gain) of biodiversity.

Finally, a number of companies implement biodiversity offsets after setting voluntary policy commitments to achieve 'no net loss' or a 'net positive impact' for biodiversity overall associated with their operations. This is part of a broader effort for the private sector to manage biodiversity.

Compensatory Mitigation in the US

No Net Loss policy (and consequently, biodiversity offsetting) has its origin in US legislation, specifically in the Water Act from the 1970s. This piece of legislation required 'no net loss of wetland acreage and function', leading eventually to the creation of mitigation banks, where wetland credits are bought and sold.

The US also has a Conservation Banking policy in which credits representing areas of habitat for protected fauna species are traded.

In the US, offsetting tends to be called 'compensatory mitigation'.

Offsetting in Australia

Biodiversity offset policies have become established in a number of Australian states (especially Victoria and New South Wales), and there is also a federal biodiversity offset policy. States tend to operate biodiversity banking mechanisms at the regional level.

Much of the scientific research into biodiversity offsetting outside of the US has been conducted by Australia, especially organisations such as CEED and CSIRO.

Offsetting in the UK

In the UK, compensation (for environmental harm caused by development) in the form of biodiversity offsetting is currently an optional (non-compulsory) tool for developers. Those developers choosing to incorporate biodiversity offsetting practices into their project plans can do so once the normal planning mitigation hierarchy has been followed, which involves taking steps to avoid and reduce environmental harm, where possible, at the development, or 'impact', site.

Biodiversity offsetting is only applicable to land that has been approved for development, which means it does not apply to protected sites such as Sites of Special Scientific Interest (SSSIs) or national nature reserves (NNRs). In addition to protected areas, vulnerable or irreplaceable habitats (such as ancient woodland) are also exempt from biodiversity offsetting.

In 2011, six biodiversity offsetting pilot schemes were started in England by the British Government to test the process. They were run in partnership with local groups and private companies and are located in Warwickshire, Essex, the Ribble Valley, at three sites in Devon, in Nottinghamshire. and Doncaster (http://www.doncaster.gov.uk/services/planning/biodiversity-offsetting-in-doncaster).

In September 2013, the British Government published a Green Paper containing plans for further incorporation of biodiversity offsetting in the UK planning system. (Public consultation period: 5 September – 7 November 2013).

Economic Value

Biodiversity is increasingly seen as having economic value due to growing recognition of the world's finite natural resources and through the benefits of ecosystem services (nature providing clean air, food and water, natural flood defences, pollination services and recreation opportunity). Placing financial value on biodiversity has created a marketplace for retaining and restoring habitats.

Financial gain from biodiversity offsetting is brought about through the sale of conservation credits by landowners. Individuals or companies who are looking to receive financial payment in return for creating or enhancing particular wildlife habitats on their property can have their land valued in conservation credits by a biodiversity offsetting broker who will then register their credits for sale to developers looking to offset any residual impact to biodiversity from their approved developments.

Developers can also find the business of biodiversity offsetting appealing financially as the compensation payment for their project's residual biodiversity impact is handled in one agreement and the landowner receiving that payment (and therefore the habitat re-creation duties) is responsible for the biodiversity restoration and management thereafter. The cost may represent a small proportion of a developer's budget and is often outweighed by a project's long-term gains. As corporate social responsibility is often part of larger companies' business priorities, being able to demonstrate environmentally responsible practices can be an additional incentive.

Biodiversity offsetting based upon showing the economic value of lost habitat is highly controversial. The schemes proposed for the UK have been regarded as failing to protect biodiversity and indeed leading to further losses in the prioritisation of development over conservation. The basic economics has been described by ecological economist Clive Spash as leading to the "bulldozing of biodiversity" under an approach that regards optimal species extinction as being necessary to achieve economic efficiency.

Conservation Credits

The cost of re-creating an area of habitat affected by a development proposal (impact site) can be calculated and represented as a number of conservation credits that a developer could purchase in order to offset their biodiversity impact. Land put forward for investment to re-create impacted biodiversity (receptor site) is also calculated in conservation credits (to account for the cost of creating or restoring biodiversity at that particular site and to cover the cost of its long-term conservation management). This situation enables the buying (by developers) and selling (by landowners) of conservation credits. Government approved (quantitative and qualitative) metrics should be used to calculate the number of conservation credits that can be applied to each site, in order to maintain accuracy and consistency in the value of a conservation credit.

Motivation

A decline in global biodiversity due, in part, to land use changes is the motivation for creating a system within the planning process that tackles unavoidable and residual impact to biodiversity. Formal evaluation of impact to habitat, wildlife and other natural considerations is often required of developers ahead of receiving approval for a project to go ahead. This can often be in the form of Environmental Impact Assessments (EIA), which are commonplace within the work of Government planning authorities. EIAs look at how proposed projects may impact upon the environment in its broadest sense, covering the traditional 'green' aspects alongside any social and economic issues; and can result in mitigating and compensatory packages which form part of a project's overall proposal for approval. The topic of biodiversity is likely to be looked at as part of an EIA, but in conjunction with many other overriding elements. Biodiversity offsetting, as an assessment and compensatory process, can either sit inside or outside of EIA and aims, specifically, to tackle habitat – and therefore biodiversity – loss.

Mitigation Banking

Mitigation banking is the preservation, enhancement, restoration or creation (PERC) of a wetland, stream, or habitat conservation area which offsets, or compensates for, expected adverse impacts to similar nearby ecosystems. The goal is to replace the exact function and value of specific habitats (i.e. biodiversity, or other ecosystem services that would be adversely affected by a proposed activity or project. The public interest is served when enforcement agencies require more habitat as mitigation, often referred to as a mitigation ratio, than is adversely impacted by management or development of nearby acreage.

Mitigation Banking Around the World

United States

In the United States, federal agencies (under section 404 of the Clean Water Act), as well as many state and local governments, require mitigation for the disturbance or destruction of wetland, stream, or endangered species habitat. Once approved by regulatory agencies, a mitigation bank may sell credits to developers whose projects will impact these various ecosystems.

Credits are units of exchange defined as the ecological value associated with converting to other economic uses a naturally occurring wetland or other specific habitat type. Mitigation credits to compensate for riparian impacts may be assigned in relation to the linear distance of a stream functioning at the highest possible capacity within the watershed of the bank.

Credits are designated by an interagency Mitigation Bank Review Team (MBRT). The MBRT evaluates and permits a proposed Mitigation Bank. The MBRT may include representatives of various federal, state and/or local government agencies, including: U.S. Army Corps of Engineers, National Marine Fisheries Service, Environmental Protection Agency, US Fish and Wildlife Service, State Environmental Protection Divisions, Local Water Management Districts, County Environmental Departments and the Soil Conservation Service.

Advantages

There are several advantages to drawing on mitigation bank credits. For example:

- Mitigation banks place a perpetual conservation easement on the land, with a trust fund specifically dedicated to long term management of natural resources inherent to the bank. By securing mitigation credits from neighboring ecosystems many large landowners, including the government, are able to maintain a property in its current management state (e.g. grazing, timber removal, low-impact recreation or education) while retaining ecological functionality, also called ecosystem services, important to the public interest.

- Mitigation banks usually provide greater benefits than on-site or small parcel mitigation efforts. Landowners PERC properties that are of higher ecological quality than the small parcel impacts they compensate for. By consolidating activities necessary to create, maintain, and monitor mitigation, banks are able to provide superior ecosystem services at a reduced cost.

- Mitigation banks provide many functional business advantages, allowing for ease of development. They allow a developer to maximize the use of a preferred development site rather than breaking up the site into sub-optimal property uses. Because mitigation bank credits are negotiated prior to development, hence prior to impact, purchasing credits from a mitigation bank decreases permitting time while also ensuring no net loss of habitat. The cost is often lower than other alternatives. Both regulatory and long term management risk is passed from developer to mitigation banker.

Despite policies mandating no net loss of habitat value and function, agencies have had difficulty ensuring that mitigation programs are managed to this outcome. Wetlands mitigation programs, for example, have in some cases been approved based on total numbers of acres rather than in terms of equivalence in ecological value or function. Merely assuming that the compensation involves a similar number of acres falls short of true equivalence unless the replacement ecological functions supplied by those acres are also the same.

References

- European Environment Agency Protected areas in Europe – an overview In: EEA Report No 5/2012 Kopenhagen: 2012 ISBN 978-92-9213-329-0 ISSN 1725-9177.

- McMillan, A.J.S.; Horobin, J.F. (1995), Christmas Cacti : The genus Schlumbergera and its hybrids (p/b ed.), Sherbourne, Dorset: David Hunt, ISBN 978-0-9517234-6-3

- ten Kate, Kerry; Bishop, J.; Bayon, R. (2004), "Biodiversity offsets, views, experience and the business case" (PDF), International Union for Conservation of Nature, Gland, Switzerland, and Insight Investment, London, ISBN 2-8317-0854-0, retrieved 4 January 2014

- Tom Regan (2006). Jaulas Vacías. El Desafío de los Derechos de los Animales (in Spanish). Barcelona: Fundación Altarriba. p. 111. ISBN 978-84-611-0672-1

- "National parks". Department of Communications, Information Technology and the Arts. Australian Government. 31 July 2007. Retrieved 2 November 2014.

- U.S. Office of Personnel Management. Handbook of occupational groups and families. Washington, D.C. January 2008. Page 19. OPM.gov Accessed November 2, 2014.

- Gissibl, B., S. Höhler and P. Kupper, 2012, Civilizing Nature, National Parks in Global Historical Perspective, Berghahn, Oxford

Global Initiatives and Policies

On a global scale, there are several organizations in place to help safeguard biodiversity. There are also many treaties and agreements in place like the Convention on Biological Diversity, Cartagena protocol on Biosafety, Nagoya Protocol and the Biodiversity Indicators Partnership. This chapter comprehensively details all these initiatives and informs the reader about the objectives of each.

Convention on Biological Diversity

The Convention on Biological Diversity (CBD), known informally as the Biodiversity Convention, is a multilateral treaty. The Convention has three main goals:

1. conservation of biological diversity (or biodiversity);

2. sustainable use of its components; and

3. fair and equitable sharing of benefits arising from genetic resources

In other words, its objective is to develop national strategies for the conservation and sustainable use of biological diversity. It is often seen as the key document regarding sustainable development.

The Convention was opened for signature at the Earth Summit in Rio de Janeiro on 5 June 1992 and entered into force on 29 December 1993.

At the 2010 10th Conference of Parties (COP) to the Convention on Biological Diversity in October in Nagoya, Japan, the Nagoya Protocol was adopted.

About

The convention recognized for the first time in international law that the conservation of biological diversity is "a common concern of humankind" and is an integral part of the development process. The agreement covers all ecosystems, species, and genetic resources. It links traditional conservation efforts to the economic goal of using biological resources sustainably. It sets principles for the fair and equitable sharing of the benefits arising from the use of genetic resources, notably those destined for commercial use. It also covers the rapidly expanding field of biotechnology through its Cartagena Protocol on Biosafety, addressing technology development and transfer, benefit-sharing and biosafety issues. Importantly, the Convention is legally binding; countries that join it ('Parties') are obliged to implement its provisions.

The convention reminds decision-makers that natural resources are not infinite and sets out a philosophy of sustainable use. While past conservation efforts were aimed at protecting particular species and habitats, the Convention recognizes that ecosystems, species and genes must be used for the benefit of humans. However, this should be done in a way and at a rate that does not lead to the long-term decline of biological diversity.

The convention also offers decision-makers guidance based on the precautionary principle that where there is a threat of significant reduction or loss of biological diversity, lack of full scientific certainty should not be used as a reason for postponing measures to avoid or minimize such a threat. The Convention acknowledges that substantial investments are required to conserve biological diversity. It argues, however, that conservation will bring us significant environmental, economic and social benefits in return.

The Convention on Biological Diversity of 2010 would ban some forms of geoengineering.

Issues

Some of the many issues dealt with under the convention include:

- Measures the incentives for the conservation and sustainable use of biological diversity.

- Regulated access to genetic resources and traditional knowledge, including Prior Informed Consent of the party providing resources.

- Sharing, in a fair and equitable way, the results of research and development and the benefits arising from the commercial and other utilization of genetic resources with the Contracting Party providing such resources (governments and/or local communities that provided the traditional knowledge or biodiversity resources utilized).

- Access to and transfer of technology, including biotechnology, to the governments and/or local communities that provided traditional knowledge and/or biodiversity resources.

- Technical and scientific cooperation.

- Coordination of a global directory of taxonomic expertise (Global Taxonomy Initiative).

- Impact assessment.

- Education and public awareness.

- Provision of financial resources.

- National reporting on efforts to implement treaty commitments.

Cartagena Protocol

The Cartagena Protocol on Biosafety of the Convention, also known as the Biosafety Protocol, was adopted in January 2000. The Biosafety Protocol seeks to protect biological diversity from the potential risks posed by living modified organisms resulting from modern biotechnology.

The Biosafety Protocol makes clear that products from new technologies must be based on the precautionary principle and allow developing nations to balance public health against economic benefits. It will for example let countries ban imports of a genetically modified organism if they feel there is not enough scientific evidence the product is safe and requires exporters to label shipments containing genetically modified commodities such as corn or cotton.

The required number of 50 instruments of ratification/accession/approval/acceptance by countries was reached in May 2003. In accordance with the provisions of its Article 37, the Protocol entered into force on 11 September 2003.

Global Strategy for Plant Conservation

In April 2002, the parties of the UN CBD adopted the recommendations of the Gran Canaria Declaration Calling for a Global Plant Conservation Strategy, and adopted a 16-point plan aiming to slow the rate of plant extinctions around the world by 2010.

Parties

As of 2016, the Convention has 196 parties, which includes 195 states and the European Union. All UN member states—with the exception of the United States—have ratified the treaty. Non-UN member states that have ratified are the Cook Islands, Niue, and the State of Palestine. The Holy See and the states with limited recognition are non-parties. The US has signed but not ratified the treaty, and has not announced plans to ratify it.

International Bodies Established

Conference of the parties: The convention's governing body is the Conference of the parties (COP), consisting of all governments (and regional economic integration organizations) that have ratified the treaty. This ultimate authority reviews progress under the Convention, identifies new priorities, and sets work plans for members. The COP can also make amendments to the Convention, create expert advisory bodies, review progress reports by member nations, and collaborate with other international organizations and agreements.

The Conference of the Parties uses expertise and support from several other bodies that are established by the Convention. In addition to committees or mechanisms established on an ad hoc basis, two main organs are:

Secretariat: The CBD Secretariat, based in Montreal, operates under the United Nations Environment Programme. Its main functions are to organize meetings, draft documents, assist member governments in the implementation of the programme of work, coordinate with other international organizations, and collect and disseminate information.

Subsidiary body for Scientific, Technical and Technological Advice (SBSTTA): The Subsidiary Body on Scientific, Technical and Technological Advice (SBSTTA). The SBSTTA is a committee composed of experts from member governments competent in relevant fields. It plays a key role in making recommendations to the COP on scientific and technical issues. 13th Meeting of the Subsidiary Body on Scientific, Technical and Technological Advice (SBSTTA-13) held from 18 to 22 February 2008 in the Food and Agriculture Organization at Rome, Italy. SBSTTA-13 delegates met in the Committee of the Whole in the morning to finalize and adopt recommendations on the in-depth reviews of the work programmes on agricultural and forest biodiversity and SBSTTA's modus operandi for the consideration of new and emerging issues. The closing plenary convened in the afternoon to adopt recommendations on inland waters biodiversity, marine biodiversity, invasive alien species and biodiversity and climate change. The current chairperson of the SBSTTA is Dr. Senka Barudanovic.

Country Implementation

National Biodiversity Strategies and Action Plans (NBSAP)

"National Biodiversity Strategies and Action Plans (NBSAPs) are the principal instruments for implementing the Convention at the national level (Article 6). The Convention requires countries to prepare a national biodiversity strategy (or equivalent instrument) and to ensure that this strategy is mainstreamed into the planning and activities of all those sectors whose activities can have an impact (positive and negative) on biodiversity. To date [2012-02-01], 173 Parties have developed NBSAPs in line with Article 6."

For example, the United Kingdom, New Zealand and Tanzania have carried out elaborate responses to conserve individual species and specific habitats. The United States of America, a signatory who has not yet ratified the treaty, has produced one of the most thorough implementation programs through species Recovery Programs and other mechanisms long in place in the USA for species conservation.

Singapore has also established a detailed National Biodiversity Strategy and Action Plan. The National Biodiversity Centre of Singapore represents Singapore in the Convention for Biological Diversity.

National Reports

In accordance with Article 26 of the Convention, Parties prepare national reports on the status of implementation of the Convention.

Executive Secretary

The current executive secretary is Braulio Ferreira de Souza Dias, who took up this post on 15 February 2012. Dr Ahmed Djoghlaf was the previous executive secretary.

Nagoya Protocol

The Nagoya Protocol on Access to Genetic Resources and the Fair and Equitable Sharing of Benefits Arising from their Utilization to the Convention on Biological Diversity is a supplementary agreement to the Convention on Biological Diversity. It provides a transparent legal framework for the effective implementation of one of the three objectives of the CBD: the fair and equitable sharing of benefits arising out of the utilization of genetic resources. The Protocol was adopted on 29 October 2010 in Nagoya, Aichi Province, Japan, and entered into force on 12 October 2014. Its objective is the fair and equitable sharing of benefits arising from the utilization of genetic resources, thereby contributing to the conservation and sustainable use of biodiversity.

Meetings of the Parties

1994 COP 1

The first ordinary meeting of the parties to the convention took place in November and December 1994, in Nassau, Bahamas.

1995 COP 2

The second ordinary meeting of the parties to the convention took place in November 1995, in Jakarta, Indonesia.

1996 COP 3

The third ordinary meeting of the parties to the convention took place in November 1996, in Buenos Aires, Argentina.

1998 COP 4

The fourth ordinary meeting of the parties to the convention took place in May 1998, in Bratislava, Slovakia.

1999 EXCOP 1

The First Extraordinary Meeting of the Conference of the Parties took place in February 1999, in Cartagena, Colombia.

2000 COP 5

The fifth ordinary meeting of the parties to the convention took place in May 2000, in Nairobi, Kenya.

2002 COP 6

The sixth ordinary meeting of the parties to the convention took place in April 2002, in The Hague, Netherlands.

2004 COP 7

The seventh ordinary meeting of the parties to the convention took place in February 2004, in Kuala Lumpur, Malaysia.

2006 COP 8

The eighth ordinary meeting of the parties to the convention took place in March 2006, in Curitiba, Brazil.

2008 COP 9

The ninth ordinary meeting of the parties to the convention took place in May 2008, in Bonn, Germany.

2010 COP 10

The tenth ordinary meeting of the parties to the convention took place in October 2010, in Nagoya, Japan.

2012 COP 11

Leading up to the Conference of the Parties (COP 11) meeting on biodiversity in Hyderabad, India 2012, preparations for a World Wide Views on Biodiversity has begun, involving old and new partners and building on the experiences from the World Wide Views on Global Warming.

2014 COP 12

Under the theme, "Biodiversity for Sustainable Development," thousands of representatives of governments, NGOs, indigenous peoples, scientists and the private sector gathered in Pyeongchang, Republic of Korea in October 2014 for the 12th meeting of the Conference of the Parties to the Convention on Biological Diversity (COP 12). http://www.cbd.int/cop2014; Webcasting: http://www.liveto.com/cop12/floor/index.html

From 6–17 October 2014, Parties discussed the implementation of the Strategic Plan for Biodiversity 2011-2020 and its Aichi Biodiversity Targets, which are to be achieved by the end of this decade. The results of Global Biodiversity Outlook 4, the flagship assessment report of the CBD informed the discussions.

The conference gave a mid-term evaluation to the UN Decade on Biodiversity (2011-2020) initiative, which aims to promote the conservation and sustainable use of nature.

At the end of the meeting, the meeting adopted the "Pyeongchang Road Map," which addresses ways to achieve biodiversity through technology cooperation, funding and strengthening the capacity of developing countries. (Source http://www.cbd.int/doc/press/2014/pr-2014-10-06-cop-12-en.pdf)

2016 COP 13

The thirteenth ordinary meeting of the parties to the convention takes will take place December 2016 in Cancun, Mexico

Commemorative Periods

2010 was the International Year of Biodiversity. The Secretariat of the Convention on Biological Diversity is the focal point for the International Year of Biodiversity. On 22 December 2010, the UN declared the period from 2011 to 2020 as the United Nations Decade on Biodiversity. They, hence, followed a recommendation of the CBD signatories during COP10 at Nagoya in October 2010.

Criticism

Although the convention explicitly states that all forms of life are covered by its provisions, examination of reports and of national biodiversity strategies and action plans submitted by participating countries shows that in practice this is not happening. The fifth report of the European Union, for example, makes frequent reference to animals (particularly fish) and plants, but does not mention bacteria, fungi or protists at all. The International Society for Fungal Conservation has assessed more than 100 of these CBD documents for their coverage of fungi using defined criteria to place each in one of six categories. No documents were assessed as good or adequate, less than 10% as nearly adequate or poor, and the rest as deficient, seriously deficient or totally deficient.

Cartagena Protocol on Biosafety

The Cartagena Protocol on Biosafety to the Convention on Biological Diversity is an international agreement on biosafety as a supplement to the Convention on Biological Diversity effective since 2003. The Biosafety Protocol seeks to protect biological diversity from the potential risks posed by genetically modified organisms resulting from modern biotechnology.

The Biosafety Protocol makes clear that products from new technologies must be based on the precautionary principle and allow developing nations to balance public health against economic benefits. It will for example let countries ban imports of genetically modified organisms if they feel there is not enough scientific evidence that the product is safe and requires exporters to label shipments containing genetically altered commodities such as corn or cotton.

The required number of 50 instruments of ratification/accession/approval/acceptance by countries was reached in May 2003. In accordance with the provisions of its Article 37, the Protocol entered into force on 11 September 2003. As of March 2015, the Protocol had 170 parties, which includes 167 United Nations member states, the State of Palestine, and the European Union.

Objective

In accordance with the precautionary approach, contained in Principle 15 of the Rio Declaration on Environment and Development, the objective of the Protocol is to contribute to ensuring an adequate level of protection in the field of the safe transfer, handling and use of 'living modified organisms resulting from modern biotechnology' that may have adverse effects on the conservation and sustainable use of biological diversity, taking also into account risks to human health, and specifically focusing on transboundary movements (Article 1 of the Protocol, SCBD 2000).

Living Modified Organisms (LMOs)

The protocol defines a 'living modified organism' as any living organism that possesses a novel combination of genetic material obtained through the use of modern biotechnology, and 'living organism' means any biological entity capable of transferring or replicating genetic material, including sterile organisms, viruses and viroids. 'Modern biotechnology' is defined in the Protocol to mean the application of in vitro nucleic acid techniques, or fusion of cells beyond the taxonomic family, that overcome natural physiological reproductive or recombination barriers and are not techniques used in traditional breeding and selection. 'Living modified organism (LMO) Products' are defined as processed material that are of living modified organism origin, containing detectable novel combinations of replicable genetic material obtained through the use of modern biotechnology (for instance, flour from GM maize). 'Living modified organism intended for direct use as food or feed, or for processing (LMO-FFP)' are agricultural commodities from GM crops. Overall the term 'living modified organisms' is equivalent to genetically modified organism – the Protocol did not make any distinction between these terms and did not use the term 'genetically modified organism.'

Precautionary Approach

One of the outcomes of the United Nations Conference on Environment and Development (also known as the Earth Summit) held in Rio de Janeiro, Brazil, in June 1992, was the adoption of the Rio Declaration on Environment and Development, which contains 27 principles to underpin sustainable development. Commonly known as the precautionary principle, Principle 15 states that "In order to protect the environment, the precautionary approach shall be widely applied by States according to their capabilities. Where there are threats of serious or irreversible damage, lack of full scientific certainty shall not be used as a reason for postponing cost-effective measures to prevent environmental degradation."

Elements of the precautionary approach are reflected in a number of the provisions of the Protocol, such as:

- The preamble, reaffirming "the precautionary approach contained in Principle 15 of the Rio Declaration on environment and Development";

- Article 1, indicating that the objective of the Protocol is "in accordance with the precautionary approach contained in Principle 15 of the Rio Declaration on Environment and Development";

- Article 10.6 and 11.8, which states "Lack of scientific certainty due to insufficient relevant scientific information and knowledge regarding the extent of the potential adverse effects of an LMO on biodiversity, taking into account risks to human health, shall not prevent a Party of import from taking a decision, as appropriate, with regard to the import of the LMO in question, in order to avoid or minimize such potential adverse effects."; and

- Annex III on risk assessment, which notes that "Lack of scientific knowledge or scientific consensus should not necessarily be interpreted as indicating a particular level of risk, an absence of risk, or an acceptable risk."

Application

The Protocol applies to the transboundary movement, transit, handling and use of all living modified organisms that may have adverse effects on the conservation and sustainable use of biological diversity, taking also into account risks to human health (Article 4 of the Protocol, SCBD 2000).

Parties and Non-parties

The governing body of the Protocol is called the Conference of the Parties to the Convention serving as the meeting of the Parties to the Protocol (also the COP-MOP). The main function of this body is to review the implementation of the Protocol and make decisions necessary to promote its effective operation. Decisions under the Protocol can only be taken by Parties to the Protocol. Parties to the Convention that are not Parties to the Protocol may only participate as observers in the proceedings of meetings of the COP-MOP.

The Protocol addresses the obligations of Parties in relation to the transboundary movements of LMOs to and from non-Parties to the Protocol. The transboundary movements between Parties and non-Parties must be carried out in a manner that is consistent with the objective of the Protocol. Parties are required to encourage non-Parties to adhere to the Protocol and to contribute information to the Biosafety Clearing-House.

Relationship with the WTO

A number of agreements under the World Trade Organization (WTO), such as the Agreement on the Application of Sanitary and Phytosanitary Measures (SPS Agreement) and the Agreement on Technical Barriers to Trade (TBT Agreement), and the Agreement on Trade-Related Aspects of Intellectual Property Rights (TRIPs), contain provisions that are relevant to the Protocol. The Protocol states in its preamble that parties:

- Recognize that trade and environment agreements should be mutually supportive;

- Emphasize that the Protocol is not interpreted as implying a change in the rights and obligations under any existing agreements; and

- Understand that the above recital is not intended to subordinate the Protocol to other international agreements.

Main Features

Overview of Features

The Protocol promotes biosafety by establishing rules and procedures for the safe transfer, handling, and use of LMOs, with specific focus on transboundary movements of LMOs. It features a set of procedures including one for LMOs that are to be intentionally introduced into the environment called the advance informed agreement procedure, and one for LMOs that are intended to be used directly as food or feed or for processing. Parties to the Protocol must ensure that LMOs are handled, packaged and transported under conditions of safety. Furthermore, the shipment of LMOs subject to transboundary movement must be accompanied by appropriate documentation specifying, among other things, identity of LMOs and contact point for further information. These procedures and requirements are designed to provide importing Parties with the necessary information needed for making informed decisions about whether or not to accept LMO imports and for handling them in a safe manner.

The Party of import makes its decisions in accordance with scientifically sound risk assessments. The Protocol sets out principles and methodologies on how to conduct a risk assessment. In case of insufficient relevant scientific information and knowledge, the Party of import may use precaution in making their decisions on import. Parties may also take into account, consistent with their international obligations, socio-economic considerations in reaching decisions on import of LMOs.

Parties must also adopt measures for managing any risks identified by the risk assessment, and they must take necessary steps in the event of accidental release of LMOs.

To facilitate its implementation, the Protocol establishes a Biosafety Clearing-House for Parties to exchange information, and contains a number of important provisions, including capacity-building, a financial mechanism, compliance procedures, and requirements for public awareness and participation.

Procedures for Moving LMOs Across Borders

Advance Informed Agreement

The "Advance Informed Agreement" (AIA) procedure applies to the first intentional transboundary movement of LMOs for intentional introduction into the environment of the Party of import. It includes four components: notification by the Party of export or the exporter, acknowledgment of receipt of notification by the Party of import, the decision procedure, and opportunity for review of decisions. The purpose of this procedure is to ensure that importing countries have both the opportunity and the capacity to assess risks that may be associated with the LMO before agreeing to its import. The Party of import must indicate the reasons on which its decisions are based (unless consent is unconditional). A Party of import may, at any time, in light of new scientific information, review and change a decision. A Party of export or a notifier may also request the Party of import to review its decisions.

However, the Protocol's AIA procedure does not apply to certain categories of LMOs:

- LMOs in transit;

- LMOs destined for contained use;

- LMOs intended for direct use as food or feed or for processing

While the Protocol's AIA procedure does not apply to certain categories of LMOs, Parties have the right to regulate the importation on the basis of domestic legislation. There are also allowances in the Protocol to declare certain LMOs exempt from application of the AIA procedure.

LMOs Intended for Food or Feed, or for Processing

LMOs intended for direct use as food or feed, or processing (LMOs-FFP) represent a large category of agricultural commodities. The Protocol, instead of using the AIA procedure, establishes a more simplified procedure for the transboundary movement of LMOs-FFP. Under this procedure, A Party must inform other Parties through the Biosafety Clearing-House, within 15 days, of its decision regarding domestic use of LMOs that may be subject to transboundary movement.

Decisions by the Party of import on whether or not to accept the import of LMOs-FFP are taken under its domestic regulatory framework that is consistent with the objective of the Protocol. A developing country Party or a Party with an economy in transition may, in the absence of a domestic regulatory framework, declare through the Biosafety Clearing-House that its decisions on the first import of LMOs-FFP will be taken in accordance with risk assessment as set out in the Protocol and time frame for decision-making.

Handling, Transport, Packaging and Identification

The Protocol provides for practical requirements that are deemed to contribute to the safe movement of LMOs. Parties are required to take measures for the safe handling, packaging and transportation of LMOs that are subject to transboundary movement. The Protocol specifies requirements on identification by setting out what information must be provided in documentation that should accompany transboundary shipments of LMOs. It also leaves room for possible future development of standards for handling, packaging, transport and identification of LMOs by the meeting of the Parties to the Protocol.

Each Party is required to take measures ensuring that LMOs subject to intentional transboundary movement are accompanied by documentation identifying the LMOs and providing contact details of persons responsible for such movement. The details of these requirements vary according to the intended use of the LMOs, and, in the case of LMOs for food, feed or for processing, they should be further addressed by the governing body of the Protocol. (Article 18 of the Protocol, SCBD 2000).

The first meeting of the Parties adopted decisions outlining identification requirements for different categories of LMOs (Decision BS-I/6, SCBD 2004). However, the second meeting of the Parties failed to reach agreement on the detailed requirements to identify LMOs intended for direct use as food, feed or for processing and will need to reconsider this issue at its third meeting in March 2006.

Biosafety Clearing-House

The Protocol established a Biosafety Clearing-House (BCH), in order to facilitate the exchange of scientific, technical, environmental and legal information on, and experience with, living modified organisms; and to assist Parties to implement the Protocol (Article 20 of the Protocol, SCBD 2000). It was established in a phased manner, and the first meeting of the Parties approved the transition from the pilot phase to the fully operational phase, and adopted modalities for its operations (Decision BS-I/3, SCBD 2004).

Nagoya Protocol

The Nagoya Protocol on Access to Genetic Resources and the Fair and Equitable Sharing of Benefits Arising from their Utilization to the Convention on Biological Diversity, also known as the Nagoya Protocol on Access and Benefit Sharing (ABS) is a 2010 supplementary agreement to the 1992 Convention on Biological Diversity (CBD). Its aim is the implementation of one of the three objectives of the CBD: the fair and equitable sharing of benefits arising out of the utilization of genetic resources, thereby contributing to the conservation and sustainable use of biodiversity. However, there are concerns that the added bureaucracy and legislation will, overall, be damaging to the monitoring and collection of biodiversity, to conservation, to the international response to infectious diseases, and to research.

The Protocol was adopted on 29 October 2010 in Nagoya, Japan, and entered into force on 12 October 2014. It has been ratified by 86 parties, which includes 85 UN member states and the European Union. It is the second Protocol to the CBD; the first is the 2000 Cartagena Protocol on Biosafety.

Scope

The Nagoya Protocol applies to genetic resources that are covered by the CBD, and to the benefits arising from their utilization. The Protocol also covers traditional knowledge associated with genetic resources that are covered by the CBD and the benefits arising from its utilization

Obligations

The Nagoya Protocol sets out obligations for its contracting parties to take measures in relation to access to genetic resources, benefit-sharing and compliance.

Access Obligations

Domestic-level access measures aim to:

- Create legal certainty, clarity and transparency

- Provide fair and non-arbitrary rules and procedures

- Establish clear rules and procedures for prior informed consent and mutually agreed terms

- Provide for issuance of a permit or equivalent when access is granted

- Create conditions to promote and encourage research contributing to biodiversity conservation and sustainable use

- Pay due regard to cases of present or imminent emergencies that threaten human, animal or plant health

- Consider the importance of genetic resources for food and agriculture for food security

Benefit-sharing Obligations

Domestic-level benefit-sharing measures aim to provide for the fair and equitable sharing of benefits arising from the utilization of genetic resources with the contracting party providing genetic resources. Utilization includes research and development on the genetic or biochemical composition of genetic resources, as well as subsequent applications and commercialization. Sharing is subject to mutually agreed terms. Benefits may be monetary or non-monetary such as royalties and the sharing of research results.

Compliance Obligations

Specific obligations to support compliance with the domestic legislation or regulatory requirements of the contracting party providing genetic resources, and contractual obligations reflected in mutually agreed terms, are a significant innovation of the Nagoya Protocol. Contracting Parties are to:

- Take measures providing that genetic resources utilized within their jurisdiction have been accessed in accordance with prior informed consent, and that mutually agreed terms have been established, as required by another contracting party

- Cooperate in cases of alleged violation of another contracting party's requirements

- Encourage contractual provisions on dispute resolution in mutually agreed terms

- Ensure an opportunity is available to seek recourse under their legal systems when disputes arise from mutually agreed terms

- Take measures regarding access to justice

- Take measures to monitor the utilization of genetic resources after they leave a country including by designating effective checkpoints at any stage of the value-chain: research, development, innovation, pre-commercialization or commercialization

Implementation

The Nagoya Protocol's success will require effective implementation at the domestic level. A range of tools and mechanisms provided by the Nagoya Protocol will assist contracting Parties including:

- Establishing national focal points (NFPs) and competent national authorities (CNAs) to serve as contact points for information, grant access or cooperate on issues of compliance

- An Access and Benefit-sharing Clearing-House to share information, such as domestic regulatory ABS requirements or information on NFPs and CNAs

- Capacity-building to support key aspects of implementation.

Based on a country's self-assessment of national needs and priorities, capacity-building may help to:

- Develop domestic ABS legislation to implement the Nagoya Protocol

- Negotiate MAT

- Develop in-country research capability and institutions

- Raise awareness

- Transfer technology

- Target financial support for capacity-building and development initiatives through the GEF

Criticism

Many scientists have voiced concern over the protocol, fearing the increased red tape will hamper disease prevention and conservation efforts, and that the threat of possible imprisonment of scientists will have a chilling effect on research. Non-commercial biodiversity researchers and institutions such as natural history museums fear maintaining biological reference collections and exchanging material between institutions will become difficult.

Biodiversity Indicators Partnership

The Biodiversity Indicators Partnership (BIP) brings together a host of international organizations working on indicator development, to provide the best available information on biodiversity trends to the global community. The Partnership was initially established to help monitor progress towards the Convention on Biological Diversity (CBD) 2010 Biodiversity target. However, since its establishment in 2006 the BIP has developed a strong identity not only within the CBD but with other Multilateral Environmental Agreements (MEAs), national and regional governments and other sectors. As a result, the Partnership will continue through international collaboration and cooperation to provide biodiversity indicator information and trends into the future.

Biodiversity Indicators Partnership Logo

Current Status

The Biodiversity Indicators Partnership is currently in a renewal phase. The Partnership is expanding in breadth and knowledge to ensure that it can play central role in a range of processes over the course of the coming decade, including supporting the United Nations Conference on Sustainable Development (Rio+20), the Millennium Development Goals (MDGs), the United Nations Convention to Combat Desertification (UNCCD), the Ramsar Convention on Wetlands, the United Nations Environment Programme and the Intergovernmental Science-Policy Platform on Biodiversity and Ecosystem Services (IPBES). Central to the renewed Partnership will be its revitalized relationship with the Convention on Biological Diversity. In 2010, at the 10th Conference of the Parties to the CBD held in Nagoya, Japan the BIP was referenced eight times in the official adopted decisions. These references demonstrated a clear will for the Partnership to continue supporting the CBD with implementation of the new Strategic Plan for Biodiversity 2011-2020. The Strategic Plan consists of 20 new biodiversity targets for 2020, termed the 'Aichi Biodiversity Targets'. Official decisions on an indicator set to measure progress towards the Aichi Biodiversity Targets will help cement the role the Partnership will play in supporting the CBD.

In the meantime a review of the existing partnership and its structure will lead to revised roles and an expanded membership to maintain and enhance the BIP's position as the leading authority on global, regional and national indicator development and production.

Background

Biodiversity encompasses the entire variety of life on Earth. It is vital for human survival and is a key measure of the health of our planet. Human activities are irreversibly impacting biodiversity. In all regions of the world species extinction rates have increased, ecosystems have been degraded, and genetic diversity has declined.

In response to this situation, the international community agreed *"to achieve by 2010 a significant reduction of the current rate of biodiversity loss at global, regional and national level as a contribution to poverty alleviation and to the benefit of all life on Earth."* This '2010 Biodiversity Target' was adopted by governments in 2002 at the 6th Conference of the Parties (COP 6) of the Convention on Biological Diversity (CBD).

An essential part of reaching the 2010 biodiversity target was being able to measure and communicate progress. For this purpose the CBD adopted a framework in 2004, which included the use of a range of biodiversity indicators to measure progress towards the 2010 target. In 2006 this framework was further elaborated and the '*2010 Biodiversity Indicators Partnership* was established, as a global initiative to further develop and promote indicators for the consistent monitoring and assessment of biodiversity. The 2010 BIP was established with major support from the Global Environment Facility (GEF).

Objectives

The main objective of the Partnership is a reduction in the rate of biodiversity loss at the global level, through improved decisions for the conservation of global biodiversity. In order to meet this objective the key outcomes of the BIP are:

(1) The generation of information on biodiversity trends which is useful to decision makers;

(2) To ensure improved global indicators are implemented and available;

(3) To establish links between biodiversity initiatives at the national and regional levels to enable capacity building and improve the delivery of the biodiversity indicators.

Biodiversity Indicators

Through CBD governance and advisory bodies, the global biodiversity community identified a suite of 17 headline indicators from the seven focal areas for assessing progress towards, and communicating the 2010 target at a global level.

Since 2007, partners have worked to ensure the most accurate information is available to decision makers. The BIP indicators have substantially contributed to the 3rd edition of the *Global Biodiversity Outlook*, featuring in the *Status and Trends in Biodiversity* chapter of this flagship CBD publication.

The Partnership also works to integrate indicator results into coherent, compelling storylines giving a more understandable picture of the status of biodiversity.

Focal areas	Headline indicators
Status and trends of the components of biodiversity	Trends in extent of selected biomes, ecosystems and habitats Trends in abundance and distribution of selected species Coverage of protected areas Change in status of threatened species Trends in genetic diversity
Sustainable Use	Proportion of products derived from sustainable sources Ecological Footprint and related concepts
Threats to Biodiversity	Nitrogen Deposition Invasive Alien Species
Ecosystem integrity and ecosystem goods and services	Marine Trophic Index Water Quality Connectivity/fragmentation of ecosystems Health and well being of communities Biodiversity for food and medicine
Status of traditional knowledge, innovations and practices	Status and trends of linguistic diversity and numbers of speakers of indigenous languages
Status of access and benefit sharing	*To be determined*
Status of resource transfers	Official development assistance provided in support of the Convention

Partners

The Partnership brings together a host of international organisations working to support the regular delivery of biodiversity indicators at the global, national and regional levels.

Partners of the BIP can be separated into the following categories:

STEERING COMMITTEE Advise on the general direction of the BIP project, and review and provide advice on key outputs	KEY INDICATOR PARTNERS Develop and implement the biodiversity indicators	ASSOCIATE INDICATOR PARTNERS Assist in the development and implementation of the indicator suite, and/or provide support to the Partnership	AFFILIATE PARTNERS Work towards similar aims and objectives as the BIP, although at different scales
Secretariat of the Convention on Biological Diversity(CBD) European Environment Agency(EEA) Food and Agriculture Organization of the United Nations (FAO) Global Environment Facility (GEF) International Union for Conservation of Nature(IUCN) Ramsar Convention on Wetlands United Nations Environment Programme (UNEP) United Nations Environment Programme World Conservation Monitoring Centre (UNEP WCMC)	Bioversity International BirdLife International Conservation International (CI) Food and Agriculture Organization of the United Nations (FAO) Global Footprint Network (GFN) Global Invasive Species Programme (GISP) Institute of Zoology, Zoological Society of London (ZSL) International Nitrogen Initiative (INI) IUCN Species Survival Commission (IUCN SSC) IUCN Sustainable Use Specialist Group IUCN World Commission on Protected Areas (IUCN WCPA) Organization for Economic Co-operation and Development (OECD) Royal Society for the Protection of Birds (RSPB) Sea Around Us Project The Nature Conservancy (TNC) TRAFFIC International Union for Ethical BioTrade (UEBT) University of British Columbia UBC Fisheries Centre United Nations Educational, Scientific and Cultural Organization (UNESCO) UNEP Global Environmental Monitoring System (GEMS) Water Programme United Nations Environment Programme World Conservation Monitoring Centre (UNEP-WCMC) University of Queensland, Australia Wetlands International World Health Organization (WHO) World Wide Fund for Nature WWF	Convention on Migratory Species (CMS) Global Biodiversity Information Facility (GBIF) International Council on Mining and Metals (ICMM) Global Land Cover Facility, NASA/ NGO Biodiversity Working Group Ramsar Convention on Wetlands Terralingua United Nations Environment Programme (UNEP) Wildlife Conservation Society (WCS)	ASEAN Centre for Biodiversity Biotrade Initiative Center for International Earth Science Information Network Countdown 2010 Circumpolar Biodiversity Monitoring Program (CBMP) Global Reporting Initiative International Indigenous Forum on Biodiversity International Centre for Integrated Mountain Development (ICIMOD) Land Degradation Assessment in Drylands (LADA) NorBio2010 PROMEBIO: A Regional Strategic Biodiversity Monitoring and Evaluation Program for Central America Streamlining European 2010 Biodiversity Indicators (SEBI2010) The Economics of Ecosystems and Biodiversity (TEEB) United Nations University - The Institute for Water, Environment and Health Tour du valat Water Footprint Network

Establishing Links

The BIP works to communicate links between the partnership's work and all potential users, including highlighting the utility of the components of the CBD indicator suite for other multilateral environmental agreements. The BIP has presented results and hosted events at major international meetings of the following MEAs: the Convention on Biological Diversity (CBD), the United Nations Convention to Combat Desertification (UNCCD), the International Treaty on Plant Genetic Resources for Food and Agriculture (ITPGRFA), the Convention on Migratory Species (CMS) and the Ramsar Convention on Wetlands. Engagement with the private sector is also an objective for the partnership. The BIP has provided financial support and currently provides ongoing technical support to the Global Reporting Initiative (GRI), to advance the integration of related indicators into corporate performance measures.

Regional and National Indicator Support and Development

In addition to improving global-scale indicators, the BIP has been actively involved in supporting national and regional initiatives; facilitating indicator development and implementation that responds to country-specific national biodiversity priorities. A programme of capacity building workshops has been run across the globe, bringing together the various institutional representatives involved in national indicator development to share experiences and best-practice. Some 45 countries have been engaged to date.

A series of guidance documents on national indicator development have been published, together with a web portal (www.bipnational.net). This multi-language website is the most comprehensive online resource of guidance, support and shared experiences of effective indicator development for countries and regions looking to develop biodiversity indicators.

Guidance Documents:

- Guidance for National Biodiversity Indicator Development and Use

- Biodiversity Indicators capacity Strengthening: experiences from Africa

- Wild Bird Index: Guidance for national and regional use

- Coverage of Protected Areas: Guidance for national and regional use

- IUCN Red List Index: Guidance for national and regional use

- Living Planet Index: Guidance for national and regional use

- Welcome to COP 10

- http://www.newscientist.com/article/mg20827854.200-what-the-un-ban-on-geoengineering-really-means.html

- *"CBD List of Parties".*

- *Hazarika, Sanjoy (23 April 1995). "India Presses U.S. to Pass Biotic Treaty". New York Times. p. 1.13.*

- *"National Biodiversity Strategies and Action Plans (NBSAPs)".*

- *Watts, Jonathan (27 October 2010). "Harrison Ford calls on US to ratify treaty on conservation". The Guardian.* London.

- http://www.cbd.int/doc/press/2012/pr-2012-02-15-Dias-en.pdf

- *"Text of the Nagoya Protocol".* cbd.int. Convention on Biological Diversity.

- *"Eighth Ordinary Meeting of the Conference of the Parties to the Convention on Biological Diversity (COP 8)".*

- Welcome to COP 9

- *"Text of the CBD".* www.cbd.int. Retrieved 7 November 2014.

- *"Fifth Report of the European Union to the Convention on Biological Diversity. June 2014"* (PDF). www.cbd.int. Retrieved 7 November 2014.

- *"The Micheli Guide to Fungal Conservation".* www.fungal-conservation.org. Retrieved 7 November 2014.

- *This article is partly based on the relevant entry in the CIA World Factbook,* as of 2008 edition.

- First signed by Kenya

- WHO 20 Questions on biotechnology

- Secretariat of the Convention on Biological Diversity (2000) *Cartagena Protocol on Biosafety to the Convention on Biological Diversity: text and annexes.* Montreal, Canada. ISBN 92-807-1924-6

- Secretariat of the Convention on Biological Diversity (2004) *Global Biosafety – From concepts to action: Decisions adopted by the first meeting of the Conference of the Parties to the Convention on Biological Diversity serving as the meeting of the Parties to the Cartagena Protocol on Biosafety.* Montreal, Canada.

- Tyrrell, T., Chenery, A., Bubb, P., Stanwell-Smith, D. & Walpole, M. (2010) Biodiversity indicators and the 2010 target: Experiences and lessons learnt from the 2010 Biodiversity Indicators Partnership. Secretariat of the Convention on Biological Diversity, Montréal, Canada. Technical Series no. 53, 196pp.

- Butchart, S.H.M., Walpole, M., Collen, B., van Strien, A., Scharlemann, J.P.W., Almond, R.E.A., Baillie, J.E.M., Bomhard, B., Brown, C., Bruno, J., Carpenter, K.E., Carr, G.M., Chanson, J., Chenery, A.M., Csirke, J., Davidson, N.C., Dentener, F., Foster, M., Galli, A., Galloway, J.N., Genovesi, P., Gregory, R.D., Hockings, M., Kapos, V., Lamarque, J.F., Leverington, F., Loh, J., McGeoch, M.A., Mcrae, L., Minasyan, A., Morcillo, M.H., Oldfield, T.E.E., Pauly, D., Quader, S., Revenga, C., Sauer, J.R., Skolnik, B., Spear, D., Stanwell-Smith, D., Stuart, S.N., Symes, A., Tierney, M., Tyrrell, T.D., Vié, J.C. & Watson, R.

2010. Global Biodiversity: Indicators of Recent Declines. Science 328: 1164-1168.

- Walpole, M., Almond, R.E.A., Besançon, C., Butchart, S.H.M., Campbell-Lendrum, D., Carr, G.M., Collen, B., Collette, L., Davidson, N.C., Dulloo, E., Fazel, A.M., Galloway, J.N., Gill, M., Goverse, T., Hockings, M., Leaman, D.J., Morgan, D.H.W., Revenga, C., Rickwood, C.J., Schutyser, F., Simons, S., Stattersfield, A.J., Tyrrell, T.D., Vié, J.C. & Zimsky, M. 2009. Tracking Progress Toward the 2010 Biodiversity Target and Beyond. Science 325(5947): 1503–1504.

- Balmford, A., Bennun, L., ten Brink, B., Cooper, D., Côté, I.M., Crane, P., Dobson, A., Dudley, N., Dutton, I., Green, R.E., Gregory, R.D., Harrison, J., Kennedy, E.T., Kremen, C., Leader-Williams, N., Lovejoy, T.E., Mace, G., May, R., Mayaux, P., Morling, P., Phillips, J., Redford, K., Ricketts, T.H., Rodríguez, J.P., Sanjayan, M., Schei, P.J., van Jaarsveld, A.S. and Walther, B.A. 2005. The Convention on Biological Diversity's 2010 Target. Science. 307(5707): 212-213. (Reprinted In: Himalayan Journal of Sciences. 3(5): 43-45).

- Dobson, A. 2005. Monitoring global rates of biodiversity change: challenges that arise in meeting the Convention on Biological Diversity (CBD) 2010 goals. Philosophical Transactions of The Royal Society B. 360 (1454): 229-241.

- Mace, G.M. and Baillie, J.E.M. 2007. The 2010 biodiversity indicators: Challenges for science and policy. Conservation Biology. 21(6): 1406-1413.

Permissions

Index